THE 80960
MICROPROCESSOR
ARCHITECTURE

THE 80960 MICROPROCESSOR ARCHITECTURE

GLENFORD J. MYERS
DAVID L. BUDDE

WILEY

A WILEY-INTERSCIENCE PUBLICATION

JOHN WILEY & SONS

NEW YORK / CHICHESTER / BRISBANE / TORONTO / SINGAPORE

Copyright © 1988 by John Wiley & Sons, Inc.

Library of Congress Cataloging in Publication Data:

Myers, Glenford J., 1946–   *B1   app 7/12/89*
    The 80960 microprocessor architecture / Glenford J. Myers, David
L. Budde.
        p.   cm.
    Includes bibliographies and index.
    ISBN 0-471-61857-8
    1. Intel 80960 (Microprocessor)   2. Computer architecture.
I. Budde, David L.   II. Title.   III. Title: Eighty nine sixty
microprocessor architecture.
QA76.8.I29284M94   1988
004.2'2—dc19   88-17240
                                                    CIP

Printed in the United States of America

10 9 8 7 6 5 4 3 2

FOREWORD

The evolution of the computer business has been intimately linked to the evolution of semiconductor technology. Increasingly powerful computers became feasible as we were able to fabricate faster, more reliable semiconductor components. Discrete semiconductor devices, the basic building blocks of early computers, gave way to integrated logic components, which led to a broad range of computer products across a price/performance spectrum. The cost of the smaller machines - the minicomputers - was low enough, and their performance adequate enough, that it became feasible to use them as computers embedded in equipment and dedicated to a well-defined set of control tasks, thus broadening the application of computer technology.

The majority of today's minicomputer and mainframe machines were architected two decades ago. Their architectures were guided by the state of the art of semiconductor technology in the 1960s. As newer generations were defined, they had to be able to run the software that already existed for the earlier machines. While this limited the flexibility to utilize fully the latest semiconductor capabilities, major improvements were still possible in the price/performance capabilities of computers, due in part to dramatic improvements in semiconductor and storage technologies. This trend continues today.

The role of the first microprocessor products was similar to that of many minicomputers: the control of a set of well-defined events. What they offered was a quantum leap in price and performance, making digital control techniques economically feasible for many applications. Although these early applications created the microcomputer business, the early machines were woefully underpowered as compared to the minicomputer and mainframe machines of their time.

As we were able to increase complexity to tens of thousands of transistors per chip, we were able to create powerful 16-bit microprocessors. Many of the embedded dedicated computers were built around these engines, usurping the role of the early minicomputers. More importantly, personal computers became a reality. Today, over ten million of these computers are being used in every walk of life - in industry, commerce, education, engineering, and many others. Microcomputer technology became the foundation of what is today the largest segment of the computer business - desktop computing.

v

But, semiconductor technology is not static; today our microprocessor products contain hundreds of thousands of transistors on a chip, and they will contain millions of transistors in the near future. How will we utilize all of that power? Clearly we will do our best to increase the power of machines that were created in the past and now enjoy wide popularity because of their large software bases, but that alone does not make full use of the technology available to us. Just like the minicomputer and mainframe designers, microcomputer designers face the dilemma that their machines were defined several technology generations ago. Nevertheless, this is an important step; it will allow us to move desktop computing to a more powerful 32-bit environment quite rapidly.

Another direction is to create new architectures based on the latest advances in computer and silicon technologies - start a new cycle with a new architecture, and do it with the knowledge that advances in silicon technology are quite predictable and that we want to take full advantage of these advances. We also need to recognize that just as happened in previous generations, many of the future applications are undefined; in our business more often than not, the power of the new technology will create a plethora of new applications. This requires that we maintain flexibility for the future.

While the above scenario puts our activities in some historical context, our job is quite pragmatic: Given that we will be able to manufacture products with millions of transistors, how do we go about utilizing this power? The task is to create the machine that can take advantage of such technology power. Given that the applicability of such machines ranges from laser printers to automotive electronics to avionics computers and countless others, many of them undefined, we need a microprocessor that can be adapted to a number of different applications. As more transistors become available to designers, we must be able to improve not only its performance, but add capabilities and incremental functions. Standards and defacto standards need to be leveraged because adherence to them will accelerate the market acceptance of the technology. And we should be able to use advances in computer technology effectively, creating more powerful machines without "topping out" too soon or having to use exotic technologies such as freon cooling to make them run fast.

This book is about the result of this work: a computer that is general purpose at its core, but tuned to a specific group of applications as a finished product. Heavy use of parallel structures, RISC-style instruction set design, and on-chip caches result in a fast core. Recognizing that some applications will require special-purpose processors (e.g., mathematical functions), or robust memory-management functions, or the use of multiple processors, and so on, the architecture accommodates the addition of such capabilities in a modular fashion around the core. Care was taken to support higher-level software in an effective manner, providing a compatible path for newer generations of products within the family.

In a way, history repeats itself; we are again designing a computer architecture with the latest silicon technology in mind. What is different is that this time we have more transistors available on the chip than many computers had in total in the not-so-distant past. The economics of the business have changed dramatically, and the applicability of the technology has broadened to the point that it is part of our everyday life; we benefited from it and became increasingly dependent on it.

It is interesting to contemplate how long it will be before compatibility with the past will become a problem in fully utilizing the latest available silicon technology. Two, three, or four generations? That is probably a safe bet. But it is hard to see much beyond that. Our architects ten years from now will probably want a clean slate to be able again to fully utilize advances we will have made in the state of the art. And that's what makes competing in this business fun - there is always a new round...

Les Vadasz
Senior Vice President
Intel Corporation

PREFACE

In the early 1980s Intel Corporation, the pioneer behind many of the innovations in the semiconductor industry, including the microprocessor, DRAM, and EPROM, made several bold decisions regarding its microprocessor business:

- to develop, despite the success of its existing 8086 architecture family, a totally new, completely incompatible, microprocessor architecture

- to develop concurrently several 32-bit microprocessors, one being the 80386, the others having this new architecture

This strategy succeeded, giving Intel two major 32-bit product lines, the 80386, a high-performance processor that is upward compatible from earlier processors, and the 80960KA, 80960KB, and 80960MC, three processors with even higher performance that are the first members of the new 80960 architecture family.

This book is about these processors and the new 80960 architecture. It does not attempt to replace Intel's reference manuals on the processors and the architecture, but supplements them by discussing the processors and the architecture from a different perspective, the perspective of *why*. For instance, when explaining aspects of the architecture, the book goes into sufficient detail to give the reader a reasonable understanding of the features being discussed, but rather than explaining them in excruciating detail, we discuss *why* the product was defined in this way. A second, but lesser, emphasis in the book is on *how*, that is, how one can use certain features of the architecture to best advantage.

We have taken this approach, which we believe to be unique for a book about a computer architecture, because we feel that understanding the *why* behind an architecture, that is, understanding what motivated the designers to choose a particular direction or add or omit a particular function, can be of tremendous benefit in understanding and using the product.

The new product line contains more than just three new microprocessors. For instance, it contains a component named the 82965 Bus Exchange Unit, which is a powerful building block for the construction of various types of multiple-processor and fault-tolerant systems. However, this book focuses, with occasional exceptions, on the processors because many more people will do software development using the 80960 product line than will do hardware development. Moreover, the processor architecture is easy to discuss independently of the particular system or application in which it is being used, but hardware system design and use of components like the 82965 can vary widely from design to design.

Chapter 1 discusses the designers' objectives, shows how the processor architecture was designed as a series of separate architectures or architecture levels, summarizes the processor architecture, and concludes with a quick tour through the internal organization of the 80960MC.

Chapters 2-4 discuss each of the three architecture levels (called core, numerics, and protected). Even though, for instance, the 80960MC microprocessor implements the protected architecture, a user of the 80960MC should not start with Chapter 4 (the chapter on the protected architecture), since each of these chapters assumes that one has read the previous chapter.

Chapter 5 discusses the bus and other external interfaces of the processor, again primarily from the point of view of *why*. Chapter 6 describes the internal organization of the processors. Although one could use the product line without ever knowing anything about the "insides" of the processors, Chapter 6 gives the reader an understanding of how different design tradeoffs were made, and additional clues about how to use the processors for maximum performance.

Chapter 7 discusses the tools and methodologies that were used to produce the 80960 and how they influenced the product. Chapter 8 is a detailed discussion of the performance of the processors, and gives software engineers and compiler designers additional insight into how to generate optimal programs. Finally, the appendix summarizes the instruction sets of the three processors.

It may be helpful in reading the book to know the authors' perspective on the subject matter. Myers initially was the manager of product-line architecture in Intel's microprocessor division, where he managed a predecessor program to the 80960 and initiated the 80386 program. He later led the 80960 architecture effort, including work on its support components. Budde started as a chip designer of the 8741 and 8748 microcontrollers and later managed the chip design of the 432 processor, during which time he developed many of the VLSI design methodologies and CAD tools used on subsequent products, such as the 80960. He later managed the chip design of the 80960 processors and, after that, got the next-generation project off the ground.

Glen Myers
Dave Budde

CONTENTS

THE 80960
MICROPROCESSOR
ARCHITECTURE

CHAPTER 1

Overview

The 80960 family consists of a new processor architecture and, for now, a first-generation of VLSI chips that implement this architecture and serve as building blocks for the construction of a wide variety of hardware system designs. The 80960 processor architecture was designed to meet the following broad objectives:

- Have a wide range of applicability, from single-chip microcontrollers to large-scale CPUs

- Be optimized for high-performance implementations

- Have high "staying" power (i.e., represent a stable software interface across multiple generations of implementations)

- Through parallel development efforts, be the optimal base for state-of-the-art operating-system and compiler technologies.

The initial VLSI chip set and the interconnect structures it provides were defined with the following objectives in mind:

- Provide a processor that is 50-100% faster than the 80386, or one that provides 6-7 "standard" MIPS at a conservative 16 MHz frequency

- Reduce system and board-area costs by providing a high level of VLSI integration

- Support a wide range of multiple-processor systems, from loosely coupled systems to tightly coupled shared-memory systems

1

•   Support a wide range of fault-tolerant capabilities, from error detection, through self-healing, to continuous operation

The first generation of 80960-family chips consists of three processors and a bus exchange unit. This generation is known as the *K series*. The second generation will consist of a series of high-speed, highly integrated microcontrollers.

MULTIPLE ARCHITECTURE LEVELS

Perhaps the most striking aspect of the 80960 processor architecture is that it is defined as three architectures instead of one, as depicted in Figure 1-1. Each level describes a complete machine, and each level is a proper subset of any and all higher levels. This means that software written for the core-architecture level, for instance, will execute correctly on an implementation of any of the higher levels.

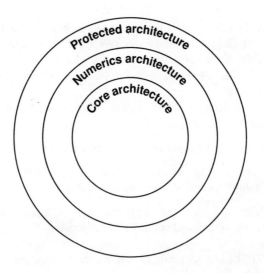

Figure 1-1:   Three levels of the processor architecture

The *core architecture* (CA) looks very much like a basic RISC (reduced instruction-set computer) machine; it defines a register-oriented instruction set, a register model, and fundamental mechanisms for such things as interrupts and faults. During the development of the overall architecture, a standalone architecture specification was developed for the core architecture, such that one could develop a processor that has only this architecture level. The 80960KA is an implementation of the core architecture.

The *numerics architecture* (NA) is an extension of the core architecture, in that it adds more computational facilities, primarily in the form of floating-point arithmetic. Instead of tacking floating-point onto the side (both architecturally and physically, as is done in most floating-point coprocessor solutions), the computational facilities added in the NA level are done so that they appear to be an integral part of the architecture. The NA level also has a complete, standalone architecture specification. The Intel 80960KB processor implements this specification.

The *protected architecture* (PA) adds two major facilities to the NA architecture: support for virtual memory, with memory protection and paging, and support for task or process management. It contains some secondary extensions as well, such as string-processing instructions and multiple-processor support. The PA level also has its own complete architecture specification, which is implemented by the Intel M80960MC.

Not shown in Figure 1-1 is a fourth level, called the *extended architecture* (XA). The XA level is a proprietary higher level of the architecture developed for use by Intel in system products; the XA level is not discussed in this book.

The rationale for structuring the architecture in four distinct, upward-compatible architectures is not easy to explain, in part because it is associated with the future evolution of the family. One of the reasons, from a chip viewpoint, is the ability to deliver an optimized cost/function ratio for different applications. For instance, a high-speed embedded controller application may need only the function of the CA or NA level, and future processors could be built specifically to these levels at a cheaper cost (and faster design time) than a processor that implements, for instance, the PA level. Of course, another alternative would have been to design specific incompatible architectures for these different levels, but from a user's point of view there is a lot to be gained in having to learn only one architecture, being able to use a single set of support tools, and being able, for instance, to execute a program for a microcontroller implementing the CA level on an 80960-based workstation implementing the PA level.

Many of the software support products, such as compilers, support the multiple architecture levels. For instance, one can inform the assembler, via an invocation switch, that the program is an NA program, in which case the assembler will flag any attempts to use instructions not defined in the NA level. If a compiler is told that the PA level is being used, the compiler will use the PA level's string instructions, but will generate an alternative code sequence if told that the CA or NA level is being used.

OTHER RELATIONSHIPS AMONG THE LEVELS

In addition to the important compatibility relationship among the architecture levels, there are other relationships of which one needs to be aware.

Beginning in Chapter 2, which explains the core-architecture (CA) level, one will note the occasional appearance of "magic values," such as the explanation that "this bit must be 1." These rules are necessary to ensure that one's software is upward-compatible to the higher architecture levels. For instance, in the PA level, there is a bit in a control register that, if set, enables address translation. In the CA and NA levels, the corresponding bit is simply labeled "must be zero." This means, for instance, that if one developed a real-time operating system for a CA or NA chip and wanted to execute it, for some reason, on a PA chip, that address translation wouldn't unexpectedly be enabled.

The significance of these magic values in the first generation of products was influenced by how these products were built. In the first generation, the three products were derived from a single modular design, and features exclusive to the 80960MC, for instance, are not necessarily absent from the 80960KB. Because of this, one can get unpredictable results in the first-generation products by violating the magic value rules (e.g., by setting, in the 80960KB, the bit that corresponds to the address-translation bit in the 80960MC). This will not be the case in second-generation products, where unique silicon designs are being done for each level.

One should also be aware that each level was defined as having certain *implementation-dependent* facilities. In other words, the architecture was defined as having a large number of facilities that will be represented in the same way from implementation to implementation, and a small number of facilities that are *not* guaranteed to be identical, or even present at all, in different implementations. Examples of the latter are how the processor is initialized, how messages are passed among processors in a multiple-processor system (because this is dependent on the external bus), and, in the 80960MC, how several instructions behave when address translation is disabled. These facilities were identified in this fashion to avoid placing unwise constraints on the current and future designs, while at the same time giving software designers fair warning about possible changes.

Finally, some aspects of the architecture were explicitly designed with future implementations in mind. For instance, many types of faults (traps or exceptions) were defined as *imprecise* to make feasible the building of future implementations that execute multiple instructions concurrently. Since the first-generation chips do not execute instructions in this way, one could write correctly functioning software for the current processors that violates this aspect of the architecture definition, but at the risk of not having the software be portable to the next implementation.

Another example is the presence of *special-function registers* (SFRs) in the architecture definition. The idea of SFRs exists, for instance, in the Intel 8096 16-bit microcontroller, where they permit such things as timers and I/O ports to be addressable as registers from any instruction. The 80960 architecture includes SFR bits in its instruction format to allow the incorporation of similar capabilities in future implementations, although the first-generation products do not define any SFRs.

THE CHIP PRODUCTS

The first generation of chip products consists of three microprocessors, the 80960KA, 80960KB, and 80960MC, and a bus exchange unit (BXU), the 82965. The M80960MC is a 32-bit processor implementing the protected-architecture level. It consists of approximately 350,000 transistors, fabricated with a 1.5-micron, two-layer metal CMOS process (the same technology used in the 80386). The chip contains an execution unit, floating-point unit, address-translation logic with a 48-entry TLB (translation lookaside buffer), and 512-byte instruction cache. The M suffix denotes that this part is tested to rigorous military specifications.

The 80960KB is similar, except that it implements the numerics-architecture level and is tested to commercial temperature specifications. Similarly, the 80960KA is a commercial temperature range implementation of the core architecture.

The processors are provided in three speed ranges: 16, 20, and 25 MHz (32, 40, and 50 MHz external clock frequencies).

All three processors have a 32-bit, multiplexed external bus. The bus is defined as a burst-transfer bus, which means that it can transfer words corresponding to successive memory locations in consecutive cycles. For instance, to fill a line in the instruction cache, the processor transmits an address and control signals indicating that the operation is a read of four 32-bit words. If the system design is a zero-wait-state design (e.g., if it uses an external cache or static RAMs), the address is transmitted in cycle 1 and the four words are returned in cycles 2-5.

Four additional pins on the processors are interrupt input signals controlling an on-chip interrupt controller. The processors also contain an output pin that signals an internal failure and an input pin that can be used to signal a bus error.

At 20 MHz and with zero wait-state operation, the processors deliver about 7.5 "standard MIPS," where one standard MIP is defined as the speed of a VAX-11/780 superminicomputer. This figure was established across a wide range of small and large benchmark programs. On the well-known Dhrystone benchmark, the performance is 14.9 million Dhrystone instructions per second.* For the well-known Whetstone floating-point benchmark, the performance is 4.2 million Whetstone instructions per second, assuming 32-bit operands. Unlike most other processors, the 80960KB and MC experience only a small loss in performance for higher preci-

---

*As a historical note, the Dhrystone benchmark was originally developed as part of the 80960 development program.

sion floating-point operands. For instance, on the 64-bit Whetstone benchmark, the performance is 4.0 million Whetstone insts/sec.

Because of certain aspects of the architecture and the chip design, such as the instruction cache, burst bus, large number of registers, and a speed-up mechanism known as the register scoreboard, the 80960 processors are much more insensitive to memory speeds than most other processors. For instance, an incremental wait state in an 80960 design costs 7-10% (in terms of speed), contrasted with 25-30% in other microprocessors.

The other component, the 82965 BXU, is oriented toward multiple-processor and fault-tolerant system designs. The BXU has two bus interfaces; one is the 32-bit bus of the processor, and the other is a different 32-bit bus known as the AP bus. The AP bus is best thought of as a backplane bus for a multiple-processor system. Its principal difference from the processor bus is that it is a transaction-oriented packet bus, meaning that read and write requests, and the replies to those requests, are treated as separate bus transactions. For instance, after a read request is transmitted on the bus (containing an address and a specification of whether 1, 2, 3, or 4 words are requested), the bus can be used to transmit additional requests before the reply to the read is returned.

In its role as an interface between the processor bus and the AP bus, the BXU component provides five major services:

- It connects multiple processors and memory modules via the AP bus.

- It serves as a cache controller and cache directory for an external cache associated with each processor. The cache controller implements a cache-coherency mechanism to provide for correct behavior when using shared data in a multiple-processor system.

- It provides a means for the interchange of messages (called *interagent communication messages*) among processors.

- By connecting 1, 2, or 4 BXUs to each processor, it supports system designs containing 1, 2, or 4 AP buses. By interleaving the physical address space over multiple buses, the aggregate bus bandwidth is increased. Also, the presence of multiple AP buses can be exploited by the BXU for recovery from bus-related failures.

- It provides an extensive set of functions for the construction of fault-tolerant systems. For instance, two BXUs can be connected in a *master/checker*[1] pair such that they check each other for failures. Two processor modules can be configured as a *primary/shadow* pair such that the passive shadow can take over for the active primary in the case of an unrecoverable error. In a multiple-bus system, the BXU can also detect unrecoverable bus errors and automatically reconfigure the system to exclude the bad bus.

Except for discussions of interagent communication messages, the BXU component is not discussed in this book. Further information on the BXU can be found in Intel's manuals on the 82965.

PROCESSOR INSTRUCTION SET

As emphasized by many advocates of RISC (reduced instruction-set computer) architectures, the encoding of machine instructions has a crucial influence on processor performance. Complex, variable-length instructions make instruction decoding the critical bottleneck in the processor by requiring several clock cycles per instruction for decoding, or by requiring an excessive number of gate delays for decoding such that the overall cycle time must be increased. Therefore, it is not surprising that the 80960 instruction-set design is similar to those of RISC machines. All instructions have a fixed size (32 bits). All instructions must be aligned on word boundaries, eliminating the need for instruction extraction and alignment logic in the critical instruction decoding path. The instruction set is register oriented; only load, store, and branch instructions have the capability to specify memory addresses.

There are five instruction formats. Most of the instructions fall into one format, the REG format, which consists of an opcode and three register designators. Each register field is six bits in width, allowing the specification of one of 32 registers and one of the commonly used constant values in the range 0-31.

This instruction encoding allows instructions to be decoded in a single clock cycle. Since the most often used instructions are executed in one cycle, and since the processors employ pipelining to overlap decoding and execution, the 25 MHz versions of the processors can execute instructions at a peak rate of one every 40 ns.

The simple and fixed-size instruction formats also allow decoding to be done in future implementations in a single cycle at much higher clock frequencies, and give future implementations the opportunity to decode multiple instructions in parallel. Considerable work went into the second-generation design, where certain classes of instructions (loads/stores, branches, and arithmetic) can be decoded and executed at a three-instruction-per-cycle rate. Given the use of multiported register arrays, the continued use of the technique of register scoreboarding, and the exploitation of smaller silicon geometries for larger on-chip data and instruction caches and frequencies of 32 MHz and above, peak execution rates of 75-100 MIPS are achievable in the next generation of 80960 implementations.

Much of the instruction set is what one would expect to find in all processors (e.g., add, multiply, shift, branch); however, some departures are summarized below.

The register-register move instructions can move 1, 2, 3, or 4 register values. The same is true for the load and store instructions (e.g., LDQ loads four words from memory into four registers). In addition to the normal shift instructions, the SHRDI (shift right dividing integer) instruction provides an adjustment to the result to allow it to be used to divide a value by a power of two (normal right-shift instructions do not divide correctly when the value is negative and odd).

All of the meaningful logical operations (e.g., and, not-and, and-not, exclusive-nor) are provided. An extensive set of bit instructions is provided (e.g., set, clear, or invert a designated bit in a register; scan for the first 0 or 1 bit; single compact instructions for testing a bit and branching if set or clear), as well as instructions for accessing bit fields. Another instruction, SCANBYTE, is particularly useful in implementing terminating strings in the C language.

All levels except for the core architecture provide an extensive set of floating-point operations, which are summarized in a later section. The protected-architecture level provides a set of instructions that operate on strings in memory (e.g., move, fill, compare).

REGISTERS

One unusual aspect of the 80960 architecture is its register model. In addition to providing a large number of registers, which are of significant benefit to optimizing compilers performing global program flow analysis and intelligent register allocation, the architecture provides multiple register banks to optimize the speed of procedure calls and returns.

At any point in time, one can address thirty-two 32-bit registers and four 80-bit floating-point registers (for extended precision floating-point values). Of the thirty-two registers, sixteen are known as global registers (named *g0-g15*) and the other sixteen are known as local registers (named *r0-r15*). The difference is that the sixteen global registers are unaffected when crossing procedure boundaries (i.e., they behave like "normal" registers in other architectures), but the local registers are affected by the call and return instructions.

When a call instruction is executed, the processor allocates a new set of 16 local registers from a pool of register sets on-chip. If the processor's pool is depleted, the processor automatically allocates a register set by taking one associated with an earlier procedure and saving it in memory. A procedure's register set is saved, if need be, in the first 16 words of that procedure's stack frame in memory. The return instruction causes the current local register set to be freed for use by a subsequent call.

Since the register sets are managed by the processor, the architecture doesn't specify the actual number of sets in the processor. The first-generation 80960s contain four sets of local registers. Remembering that every time we do a return we free a register set for a subsequent call, the processor runs out of on-chip register sets only in the situation where a program does a series of calls from a high-level procedure down to a low-level one. This certainly happens, but studies of program behavior have shown that, in terms of call/return behavior, programs spend most of their time oscillating between 2-4 levels of procedures.

In addition to reducing the overhead of saving and restoring registers when crossing procedure boundaries, the multiple sets of local registers provide three additional key performance benefits:

1. Because a local register set is saved, when needed, in the first 16 words of the stack frame in memory associated with the procedure to which the set was allocated, one can think of a local register set as a logical part of the stack frame. Therefore, with a few exceptions,* one can allocate many of a procedure's local variables directly in the local registers without allocating them in memory.

2. Because of the way in which local register sets are associated with stack frames in memory, the processor does not store linkage information in memory when a call is performed; instead it stores this information in the register sets. Specifically, three words of information are stored: the address of the previous stack frame, the stack pointer, and the instruction pointer. The first two are stored in registers r0 and r2 of the new register set and the latter is stored in register r1 of the previous register set. As a result, the processor does not need to perform any memory operations to execute call and return instructions.**

3. Because of the large number of registers, parameters should be passed in registers instead of on the stack in memory. Intel's compilers do so, using the stack only if the number of parameters exceeds the number of global registers allocated for parameter passing.

FLOATING-POINT ARITHMETIC

In a departure from most previous microprocessors, the 80960KB and 80960MC incorporate full floating-point support within the processor instead of relying on a coprocessor. Although this has important benefits in terms of cost, board space, and

---

*Exceptions are where the program does up-level referencing to the local variable, and where the program needs to take the address of the local variable, both of which are relatively uncommon.

**Except for calls where there is no available on-chip register set, and for returns where the needed register set is not on-chip.

consistency of architecture,* the principal motivation is performance; as floating-point speeds increase, the off-chip coprocessor interface becomes a significant performance bottleneck.

The floating-point part of the architecture implements the ANSI/IEEE standard 754. In addition, the architecture adds

- Suggested functions defined in the standard's appendix (e.g., copy-sign, scale, and classify functions)

- Mixed-precision operations (e.g., the ability to multiply a 32-bit operand by an 80-bit operand, with only one rounding operation)

- Transcendental functions

Three data types (of 32, 64, and 80 bits) are supported. Floating-point instructions can access data in any of the 32-bit registers (or successive registers in the case of 64- or 80-bit data types); in addition, four extra 80-bit registers are provided. The operations provided for all three data types include

- move, add, sub, mul, div, rem, sqrt, round, scale

- sin, cos, tan, atan

- $\log_2$, $\log_2(1+\varepsilon)$, logbn, 2^x-1

- compare, compare-ordered, convert, convert truncated, classify

In addition, great attention was paid in the implementation to accuracy and monotonicity.

VIRTUAL MEMORY AND PROTECTION

The 80960MC contains an on-chip memory management unit for the implementation of virtual memory via paging. The 2^{32} byte virtual address space is mapped via two levels of page tables into 4096-byte pages. The processor contains a TLB (translation lookaside buffer) that allows most virtual addresses to be translated into physical addresses in one cycle.

Two modes of privilege are defined: user mode and supervisor mode. Each page can be defined as (1) no access in user mode, read-only access in supervisor mode; (2) no access in user mode, read/write access in supervisor mode; (3) read-

---

*For instance, unlike the programming model of most coprocessors, floating-point operands can be placed in any register, and floating-point instructions use the same instruction formats, condition code, and faulting or trapping mechanism.

only access in user mode, read/write access in supervisor mode, or (4) read/write access in any mode.

Each page-table entry also contains accessed and altered flags, which are set by the processor and usable by an operating system for page replacement algorithms. Each page-table entry also contains a cacheable flag, which can be used by software to indicate to an external cache whether information in the page should be cached or not. For instance, the operating system would likely clear the cacheable flag in pages that map into parts of the physical address space into which I/O peripherals are mapped. The external bus contains a cacheable line, on which the processor puts the cacheable flag from the page associated with the memory reference.

INTERRUPTS AND FAULTS

The architecture specifies a priority interrupt mechanism and the means to remember pending interrupts. If the priority of a signaled interrupt is higher than the current processor priority, an interrupt handler is called and the processor priority is changed to that of the interrupt; otherwise, the interrupt is remembered in a pending-interrupt bit vector. Whenever the processor priority is lowered, the bit vector is checked to see if there is a pending interrupt of higher priority.

In addition to being able to accept interrupts sent as messages on the external bus and handshake with an Intel 8259A interrupt controller, the 80960 processors contain an interrupt controller with four programmable interrupt pins.

The architecture also defines a similar mechanism for the handling of *faults*, which are events associated with program execution such as arithmetic overflow, page fault, protection error, and inexact floating-point result. The architecture permits some types of faults to be *imprecise*, which enables the implementation to execute multiple instructions concurrently, or even to execute instructions out-of-order. Although this mechanism is used in only a limited way in the 80960, it exists in the architecture to allow the development of future, higher-performance processors.

PROCESS MANAGEMENT

An integral part of the protected architecture level and the 80960MC is the concept of a software process or task, which is represented by a process control block. The process control block defines a process's virtual-address space, priority, and execution mode, contains timing information for the process, and serves as a save area for the process's global registers.

The architecture provides two means for the control of process switching, known as manual and automatic process management. The former means that con-

trol is largely in the hands of a software operating system, and the latter means that
the control is largely in the hands of the chip. Manual process management is performed with two instructions, save-process and resume-process, which allow an
operating system to switch processes explicitly.

Automatic process management relies on a priority-based process scheduling
and dispatching mechanism that is built into the processor. There is a queue of processes (process control blocks) in memory from which the processor(s) will dispatch processes without any software intervention when a new process is needed
(e.g., when the current process takes some action that blocks its execution, or when
its time-slice value expires). A set of message-based and semaphore-based interprocess communication instructions is provided, which are similar to services normally provided in software operating system kernels; however in the 80960MC,
they are provided entirely in silicon.

The processor keeps track of the cumulative execution time of each process and
provides optional time-slice management. For the latter, whenever a process executes for longer than a prescribed amount of time, the processor generates a fault, or
enqueues the process on the queue of available processes and dispatches another
process.

TRACING AND DEBUGGING SUPPORT

Providing support for software debugging and tracing was another objective of
the 80960 architecture. Most of this support is invoked via a trace-controls register.
The trace controls allow detection of any combination of the following events:

- Instruction execution (i.e., single step)
- Execution of a taken branch instruction
- Execution of a call instruction
- Execution of a return instruction
- Detection that the next instruction is a return instruction
- Execution of a supervisor call
- Breakpoint (hardware breakpoint or execution of a breakpoint instruction)

When a trace event is detected, the processor generates a trace fault to give control to a software debugger or monitor. The processor contains two instruction
breakpoint registers into which a debugger can place the addresses of two instructions (e.g., for setting breakpoints in EPROM-resident code).

In addition to the above, the processor contains some proprietary debug support,
which is used by some of Intel's debugging tools and its in-circuit emulator; a special version of the processor contains some additional signal pins for use in the in-

circuit emulator. The support includes a special shadow address space in which debug software can reside and be totally invisible to users, additional event flags for trapping on process switches and resets, a mechanism to send messages about program execution flow to external hardware, and miscellaneous interrupt and signaling pins for external emulator hardware.

MULTIPROCESSING

Another important objective of the processor architecture is the support of multiple-processor systems. Although some of the support is independent of the specific configuration being used (e.g., loosely coupled with private memory, tightly coupled with shared memory), the majority of the support is oriented toward tightly coupled, shared memory multiple-processor systems, where processes can migrate from processor to processor.

The support in the processor takes two principal forms: mechanisms for processors to communicate with one another, and proper design of algorithms for correct behavior in a multiple-processor environment. The former consists of messages that can be transmitted on the external bus. That is, the processor responds to two forms of directives: instructions fetched from memory, and messages received on the bus. About twenty message types are supported, which specify such operations as interrupt, reset, stop, purge instruction cache, and purge page from TLB. A typical use is the following. If operating-system code executing on a processor makes a decision to swap a page out of memory, it must have a way of informing all other processors to purge information about this page from their TLBs. It accomplishes this by sending the appropriate message to all other processors.

Messages may also be used for other purposes. For instance, the interrupt message allows an interrupt to be directed to a processor whose priority is lower than that of the interrupt. The preemption message is used when certain designated "preempting processes" become ready to direct the process immediately to the lowest-priority processor.

The second major aspect of the multiple-processor support is the design of algorithms, principally those in the 80960MC related to process management. All pertinent algorithms use a synchronization mechanism to provide correct behavior in the event of multiple processors. For instance, the logic that provides automatic dispatching of processes from a queue first locks the queue before manipulating it. This ensures that if multiple processors attempt to dispatch simultaneously, their activities are synchronized.

INTERNAL CHIP DESIGN

The 80960 processors are implemented with a five-stage pipeline. The pipeline stages are

- fetch instruction $i+3$
- decode instruction $i+2$
- queue instruction $i+1$ for execution
- execute instruction i
- write result of instruction $i-1$

The pipeline results in a best-case execution rate of one instruction per cycle. The potential conflict between the last two pipeline stages (where the destination register of instruction $i-1$ is a source value of instruction i) is resolved by a special bypass path in the execution unit.

Because of constraints on the overall die size, the register array in the processor has only a single port, meaning that in any given cycle, only one register can be read and one register can be written. In the case of an instruction that has two input operands, this can result in the instruction taking two cycles instead of one, but this happens only infrequently. An add instruction that specifies a register as one input and a literal value (0-31) as another input takes only one cycle, because only one register value is needed. If the add instruction specifies two registers as inputs, it takes one or two cycles to execute, depending on the circumstances. If the second register is the same as the result register of the previous instruction, it takes only one cycle, because the processor uses the special bypass path mentioned above to get the second register value; otherwise it takes an extra cycle. Since an instruction such as add usually has one literal operand (e.g., adding a constant) or uses the result register of the previous instruction (e.g., in continuing a calculation), its execution time is statistically closer to one cycle than to two.

Although the processors contain a microcode ROM, most instructions are executed directly by hardware logic rather than by microcode interpretation. The main roles of the microcode logic are

- Execution of complex instructions and functions, such as string instructions, certain floating-point algorithms, process-management support, and interagent communication messages

- Handling of floating-point special cases

- Generation of faults

- Initialization

- Chip self-test

Figure 1-2 shows the overall floor plan of the 80860MC. At the center of the chip are the blocks that begin the execution pipeline: the instruction cache and instruction fetch unit. The instruction cache is a 512-byte, direct-mapped, cache with a 16-byte line size. The instruction fetch unit is responsible for initiating reads from memory when a cache miss occurs, initiating reads from memory for antici-patory prefetches of instructions into the cache, and delivering instructions from the cache to the instruction decoder.

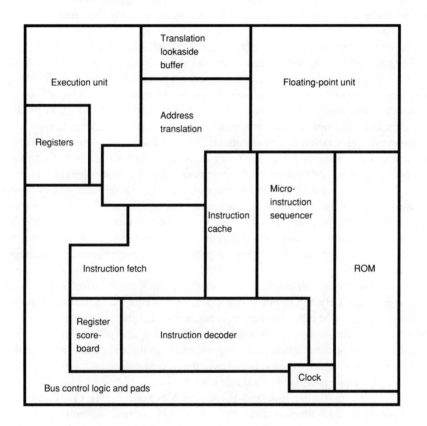

Figure 1-2: Physical organization of the 80960MC

The instruction decoder is responsible for decoding instructions and determining whether they can be delivered directly to the execution unit or the floating-point unit, or whether they need to be handled by microcode, in which case the decoder suspends instruction execution and initiates a branch into the microinstruction ROM. The instruction decoder also directly executes certain types of instructions, such as branches, and can overlap their execution with the execution of a previous instruction in one of the two execution units. To do so correctly, the instruction decoder maintains a *condition-code scoreboard* flipflop, which tells the decoder whether the previous instruction is one that will set the condition code in cases where the instruction being decoded, such as a conditional branch, is dependent on the condition code.

The instruction decoder also monitors the instruction stream with respect to the enabled trace events and calculates effective addresses for the load and store instructions.

The integer execution unit (upper left-hand corner) executes most of the register-to-register arithmetic and logical instructions. It is closely integrated with the register array, which contains 112 32-bit registers (16 global registers, 64 local registers, and 32 internal registers available for use by the microcode).

Another important block of logic is the register scoreboard, which is used to reduce the effective time of load instructions and, as shown in Chapter 8, represents an important optimization to be exploited by programmers and compilers. When a load instruction is executed, the register or registers to be loaded are marked in the scoreboard as busy, and execution of the subsequent instructions continues even as the address associated with the load instruction is placed on the external bus. If the specific registers to be used by an instruction are not marked as busy, the instruction decoder allows the instruction to proceed, but if one or more of the registers are marked as busy in the scoreboard, the instruction's execution is delayed until the registers become nonbusy (i.e., until some previous load operation into these registers completes).

The floating-point unit executes most of the floating-point instructions, as well as the integer multiply and divide instructions. It contains the four 80-bit floating-point registers, several other internal registers, a 68-bit shifter, two adders (a 68-bit adder for mantissas and a 16-bit adder for exponents), and a ROM for constants needed by some of the floating-point algorithms. The multiplication logic handles two bits per cycle, and also scans ahead over the next three and seven bits, which, if are all 0 or 1, can be processed in a single cycle.

The address-translation unit and translation lookaside buffer (TLB) translate virtual addresses into physical addresses. The TLB contains 48 entries, most of which are the most recently used page-table entries. Address translation consumes one cycle when the entry needed is already in the TLB (i.e., if a load instruction is seen in cycle i, the physical address is driven on the external bus in cycle $i+1$).

The ROM and microcode sequencer exist, as explained earlier, for implementing instructions and functions too complex to be handled directly by either of the two execution units. The ROM contains 3072 42-bit words. The microinstruction set is a superset of the instruction set of the core architecture level (i.e., there are microinstruction functions not defined in the macroarchitecture, and the microinstructions can address an additional bank of registers). Since microinstructions do not pass through the instruction decoder, and since the instruction decoder normally handles branching operations, each microinstruction contains fields to specify a branching operation within the microcode ROM.*

Finally the bus-control unit manages the external bus and most of the other external pins (e.g., interrupt pins). An important part of the bus-control logic is a queue of access buffers, which serve two purposes. First, they permit the bus-control logic to give the rest of the chip the appearance that store instructions are always executed immediately (e.g., even if they need to be queued because the external bus is busy). Second, in conjunction with the register scoreboard, they allow the processor to execute a series of load instructions without waiting for the external bus.

SIGNIFICANCE OF THE 80960

Given the number of new 32-bit microprocessor architectures that have been introduced in the marketplace, it is reasonable to wonder what's special about the 80960. In our opinion, there are six major aspects of the 80960 that differentiate it from other products.

Design for Parallelism

The best mechanism for the next leap in microprocessor performance will be the ability to execute multiple instructions in parallel. CISC (complex instruction-set computer) architectures have little potential in this area because of variable-length instructions. RISC architectures overcome this limitation, but usually contain other types of "gotcha's" that prohibit the combination of parallel instruction execution and full software compatibility with previous implementations. The 80960 architecture was defined with the objective of parallel instruction in mind, one outcome being the existence of imprecise faults.

Design for Extensibility

The 80960 is an architecture designed for a family of products, not one that started as a single product and later evolved into a family, as has been the case in other microprocessor families. For instance, the 80960 architecture definition contains some things that are not even implemented in the first generation (e.g., special function registers). Also, the architecture definition is carefully segregated

---

*Which is part of the reason why microinstructions are 10 bits wider.

into features that will remain constant in all implementations, and a small number of features (called implementation-dependent) whose representation and existence are permitted to change in future implementations.

Multiple Architecture Levels

The definition of the architecture as three distinct upward-compatible architectures gives Intel the ability to create implementations for a wide range of applications and price points, while minimizing Intel's and its customers' investments in training, development tools, and software. The ability in the future to develop and execute one's control program for an automotive anti-lock brake controller on one's workstation or minicomputer in the lab is an environment not likely to be present with any other microprocessor family.

Floating-Point Facilities

The 80960's floating-point arithmetic capabilities are significant from several points of view. First, they are fully integrated into the architecture (e.g., in terms of instructions, registers, and faults) instead of being both an architectural and physical appendage, as is the case with the floating-point coprocessors in Intel's 8086 and Motorola's 68000 families. Second, unlike other attempts to integrate floating-point functions within the processor chip, the 80960 provides a complete set of functions and data types (e.g., the 80-bit extended-real data type). Third, the facilities go beyond the IEEE standard in several worthwhile ways, such as in their ability to perform mixed length operations without incurring rounding difficulties.

IAC Messages

In a sense, the 80960 responds to two sources of instructions: a sequential program fetched from memory, and individual interagent communication messages sent to it from another source. IAC messages can be sent for execution into the bus interface of a processor from software executing on another processor and, in some circumstances, from another processor itself.

Not a Paper Tiger

At the time of initial public announcement, the 80960 processors were not products still under development, but products that had experienced considerable usage by Intel and a small group of its customers in a variety of system designs. The silicon was operational for over two years prior to public introduction. At the time of introduction, production-quality (as opposed to engineering sample) parts were available, including military-specification parts, along with a complete set of development tools.

REFERENCES

1.   G. J. Myers, A. Y. C. Yu, and D. L. House, "Microprocessor Technology Trends", *Proc. of the IEEE*, Vol. 74, No. 12, 1986, pp. 1605-1622.

CHAPTER 2

The Core Architecture

The core architecture is the innermost layer of the 80960 architecture. One can think of it as the complete architecture of a processor (the 80960KA), as well as the foundation instruction set of the numerics and protected architecture levels.

DATA TYPES

The fundamental data types of the architecture are 32-bit signed and unsigned integers, represented in two's-complement notation. Most of the instructions refer to 32-bit operands in registers. In addition, a small number of instructions can refer to register pairs, triples, or quads (64-, 96-, and 128-bit operands). Operands of load and store instructions can be 8, 16, 32, 64, 96, or 128 bits in size. The distinct set of data types defined by the instruction set is

- integer (8, 16, 32, and 64 bits)
- ordinal (unsigned integer) (8, 16, 32, and 64 bits)
- triple-word (96 bits)
- quad-word (128 bits)
- bit
- bit field (1-32 bits)

The term *byte* is used to denote an 8-bit quantity, the term *short* is used to denote a 16-bit quantity, and *long* denotes a 64-bit quantity.

Values in memory are referenced by addressing the lowest-addressed (first) byte holding part of all of the value. For numeric operands larger than a single byte, the byte containing the least-significant part of the value has the lowest memory address, and successively addressed bytes hold successively increasing parts of the value. Put more succinctly, the 80960 is a "little-endian" architecture.[1] The principal reason for doing so was for data-type consistency with the Intel 8086 architecture. The DEC VAX architecture is also little-endian; the IBM S/370 and Motorola 68000 architectures are big-endian.

When references are made to specific bits within values, the convention used is that bit 0 represents the least-significant bit, and higher bit numbers denote successively higher bits of significance. Thus, in a 32-bit integer, the sign bit is denoted as bit 31.

An area of considerable debate in recent years has been whether or not to require that operands in memory be aligned on their "natural boundaries." The natural boundary of a 32-bit operand is a 32-bit boundary (i.e., the two low-order bits of the memory address are zero); the natural boundary of a 64-bit operand is a 64-bit boundary (three low-order address bits are zero). In the early 1960s, the IBM S/360 architecture required natural boundary alignment, but this requirement was removed as the architecture evolved. Today, at least in RISC architectures, the trend is in the opposite direction (back to requiring alignment).

The rationale for required alignment is the following. First, since physical memories are addressable on only 32-bit word (or larger) boundaries, arbitrary alignment requires that the processor contain additional control logic to detect an unaligned memory operation and synthesize it out of multiple aligned memory requests. For instance, if the processor has a 32-bit memory bus and encounters a load instruction for a 32-bit word not aligned to a word boundary, the processor must issue reads of two successive words and then compose the desired result. Second, this would require multiplexer logic in the data path between the processor and its bus, and this path is likely to be a critical-speed path, meaning that additional gate delays on it can increase the processor's cycle time. Third, in architectures that provide paging, dealing with unaligned operands leads to additional complexities (e.g., dealing with the situation where an operand spans a page boundary, and one page is accessible but the other one is not).

Required operand alignment is well understood from a compiler point of view, and does not pose a problem to any of today's programming languages. In fact, languages that provide explicit means for memory addressing, such as C, carefully specify their semantics such that programs are portable between aligned and unaligned architectures. Also, since unaligned accesses are often associated with extra performance penalties in architectures that support them, many compilers simply align everything, whether required or not.

The 80960 architecture handles the issue in the following way. The core and numerics architecture levels require memory operands to be aligned to their natural boundaries; the protected architecture does not. The reason for the compromise is that, despite the above arguments, required alignment is a major obstacle in certain types of applications, such as the processing of fields in bit-mapped graphics.

Therefore, in the core architecture, 16-bit operands must be on halfword (2-byte) boundaries, 32-bit operands must be on word boundaries, 64-bit operands must be on doubleword boundaries, and 96- and 128-bit operands must be on quadword boundaries. The architecture gives the implementation the option of doing one of two things when a violation occurs: either (1) generating a fault, or (2) relaxing the above rule and performing the unaligned access. To assure portability across different 80960 implementations, one should not make any unaligned accesses, or else provide a software fault handler to perform them in software.

The first-generation chips take the second option above, and actually support unaligned accesses (although with a performance penalty, particularly when an access spans a 16-byte boundary because of the bus architecture). The reason is that these chips are based on a single chip design that implements all architecture levels, and the protected level allows misaligned operands. However, to allow portability of software to future implementations of the core architecture, one should view the initial implementation of the core architecture as requiring alignment. In fact, the second-generation implementation of the core architecture definitely requires operand alignment.

ARITHMETIC CONTROLS

Closely associated with the data types is a register containing control and status flags associated with arithmetic operations. This register, called the arithmetic controls (AC), is shown in Figure 2-1.*

The lower three bits represent the condition code, which is altered by comparison and a few other instructions, and which is tested by conditional-branch and a few other instructions. The condition code is represented in three bits (rather than fewer) to avoid having to encode and decode them. Comparisons set the condition code to 100 (less than), 010 (equal), and 001 (greater than). Other instructions set the condition code to 010 (true) or 000 (false).

---

*In the numerics-architecture level of the architecture, additional fields in the arithmetic controls are defined.

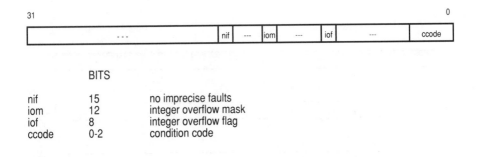

BITS		
nif	15	no imprecise faults
iom	12	integer overflow mask
iof	8	integer overflow flag
ccode	0-2	condition code

Figure 2-1: Arithmetic controls

The integer-overflow mask, if set, inhibits the generation of integer-overflow faults. If the mask is set, the integer-overflow flag is set whenever an overflow condition occurs. This flag is a "sticky" flag, meaning that it is never implicitly cleared.

INSTRUCTION FORMAT

All instructions are one word (32 bits) in length and must reside on a word boundary. The format of the instructions is shown in Figure 2-2. There are five instruction formats; the instruction's opcode defines its format.

Most of the instructions have the REG format. Instructions in this format have a 12-bit opcode (sparsely encoded opcodes reduce the decode time). In addition to the opcode, the instruction specifies three operands, denoted as *source1*, *source2*, and *src/dest*. The first two are typically source (input) values and the latter is typically the result.

The *source1* and *m1* fields define one of the instruction's source values. If *m1* = 0, *source1* (a five-bit field) refers to one of 32 registers. The encodings 0xxxx denote the local registers (r0-r15), and the encodings 1xxxx denote the global registers (g0-g15). If *m1* = 1, *source1* denotes a literal value (positive value 0-31).

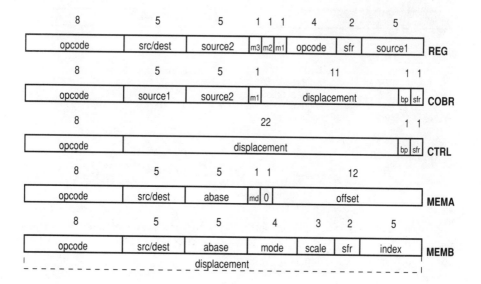

Figure 2-2:   Instruction formats

The above interpretation of *source1* and *m1* applies providing that field *sfr* is set to 0. Field *sfr* is used to extend the register space beyond 32 registers and 32 literal values. These extensions do not exist in the first implementation however, and one should set this field to 0. The *sfr* field will be used in the future to allow instructions to address special-function registers (e.g., the output port of an A/D converter, timers, I/O ports).

Two other considerations apply to the *m1* field. When the instruction defines its operand as an address rather than a data value, *m1* must be set to 0 (because *m1* = 1 in this situation is associated with a function in Intel's proprietary extended architecture). When the instruction does not use the source1 operand, one should set *m1* to 1 (i.e., encode the unused operand as an arbitrary literal value). This is because of the way the processor implements its register scoreboard; encoding an unused operand as a register could result in an unwanted scoreboard hit that slows execution of the instruction.

This also applies to fields *source2* and *m2*, which specify the second source operand of the instruction.

In a similar manner, fields *src/dest* and *m3* define the third operand. For most instructions, this operand is the result of the instruction, and *m3* should be set to 0. However, in a few cases, instructions have three source operands, in which case *src/dest* and *m3* specify an operand in the manner discussed for *source1* and *m1* above.

The rationale for allowing operands to be encoded as the literal values 0-31 was an analysis of a number of utility programs in the UNIX operating system. These programs were compiled to the VAX architecture, and the distribution of constants is shown in Table 2-1.

The table shows that the 0-31 literal encodings cover 72% of the instances. Often, the other situations can be handled without resorting to the loading of a constant from memory. For instance, the LOAD-ADDRESS instruction can move the values 0-4095 into a register, or add the value to another register. A LOAD-ADDRESS instruction followed by a ROTATE, LEFT-SHIFT, RIGHT-SHIFT, or NOT instruction can generate a wide range of 32-bit constants.

The second instruction format is the COBR (compare and branch) format. Instructions in this group have two source operands and an 11-bit signed branch displacement (in units of words). The source1 operand can be a literal or register; the source2 operand is a register. The COBR-format instructions perform a comparison operation using the two source operands and then conditionally branch based on the outcome (in which case the branch displacement is added to the instruction pointer).

Table 2-1: Distribution of constants in VAX programs

Value	Percentage
< -1	1.0
-1	3.4
0	27.8
0	27.8
1	15.9
> 31	23.7

The *bp* bit in the COBR and CTRL formats is associated with *static branch prediction*. The bit gives the compiler (or programmer) the ability to tell the processor whether or not the branch is likely to occur. Compilers can usually guess accurately in certain circumstances (e.g., for branches forming loops and for branches associated with run-time checks), and programmers may want to convey this information for branches in critical assembly-language routines. Presumably, the processor can use this information to improve the average branch time. The first-generation chips do *not* examine this bit, and the initial version of the assembler does not provide the means to set it.

The purpose of the *sfr* bit is similar to that of the *sfr* field in the REG format; it is unused in the first-generation chips, and should be encoded as 0 for compatibility with future implementations of the architecture.

All of the instructions in the COBR format are redundant, in that they represent combination of two other instructions. The COBR format provides two advantages. First, the COBR instructions improve code density (the size of programs). Second, they allow the processor to "see" an upcoming branch operation one instruction earlier and use this information to enhance the speed of branches (although the first generation does not take advantage of this).

The third format, the CTRL (control) format, contains an opcode and a 22-bit word displacement (and *bp* and *sfr* bits, which are not used in the initial processors). This format is primarily used to encode instructions that alter control flow, such as calls and branches. The signed displacement is added to the instruction pointer. The 22-bit displacement does not limit calls and branches to any arbitrary location in memory; forms of these instructions also exist in the last two forms - the MEM, or memory-addressing, formats.

LOAD, STORE, and LOAD-ADDRESS instructions (and extended forms of the CALL and BRANCH instructions) have the MEM format. These instructions specify a source or destination register (field *src/dest*) and contain other information used to compute a memory address (called the *effective address*). Actually, there are two formats: MEMA and MEMB. The opcode does not distinguish between the two; rather, bit 12 of the instruction makes the distinction.

The MEMA format specifies a base register (*abase*) and a 12-bit unsigned offset. Bit 13 specifies one of two addressing modes. If 0, the effective address is the offset (zero-extended to 32 bits); if 1, the effective address is the value of the base register plus the offset. The former mode is useful for data at absolute addresses in the bottom of memory, or for moving 12-bit constants into a register via the load-address instruction. The latter mode provides the conventional "register plus displacement" addressing mode (when the displacement is positive and can be expressed in 12 bits).

The MEMA format also would seem to provide the conventional "register indirect" addressing mode by encoding the offset as 0. Although this is the case, one should instead use the MEMB format with mode 0100, which provides the same effect but is faster (at least in the current implementations of the architecture).

The MEMB format provides additional addressing modes, which are shown in Table 2-2. The format specifies two registers that can be used in the address calculation (*abase* and *index*) and a scale factor, which is used to specify by how many bits the value from the index register is shifted.

Although the mode field is four bits in length, there are only eight encodings, because the second bit (bit 12 in the instruction) distinguishes the MEMB format from the MEMA format.

The modes that specify a displacement take the value of the next word in the instruction stream and use it as a signed 32-bit displacement. Thus some of the MEMB-format instructions are followed by a 32-bit displacement.

The most-complicated mode takes the value of the register specified by the *index* field, shifts the value left by 0-4 bits (i.e., multiplies it by 1, 2, 4, 8, or 16), adds this to the value of the register specified by the *abase* field, and adds this to the value of the following 32-bit displacement. Note that the scale factor is expressed explicitly, rather than being a function of the size of the memory operand (which is expressed in the opcode). This separation is useful when dealing with, for example, arrays of records. Consider a one-dimension array of complex numbers, where each element (a complex number) consists of two 32-bit floating-point values. If one wishes to load the real part of the *i*th element, the opcode would specify a 4-byte load, but the scale factor would be specified as 8.

Table 2-2:   MEMB addressing modes

Mode	Effective address
0100	(abase)
0101	IP + displacement + 8
0110	reserved
0111	(abase) + (index) $\times$ 2^{scale}
1100	displacement
1101	(abase) + displacement
1110	(index) $\times$ 2^{scale} + displacement
1111	(abase) + (index) $\times$ 2^{scale} + displacement

The 80960 architecture omits some traditional addressing modes, and it should be instructive to explain why. One, known as *memory deferred* or *memory indirect*, involves computing the effective address of a word in memory, which is then fetched and used as the address of the operand. This was omitted for three reasons. First, it is infrequently used (e.g., in a study of VAX programs[2], about 2% of the operand addresses are of this type), and one would expect it to be used even less in a processor with more registers and optimizing compilers. Second, it would have added considerable complexity to the instruction decoder. Third, it is slow because it requires load/store instructions to make two memory accesses. Because of the register scoreboard, one can generate faster code when a needed pointer to an operand is in memory by using two load instructions and trying to move an unrelated instruction between them.

The second missing addressing mode is *auto-increment/decrement*, where, in addition to computing an effective address, the index register is incremented or decremented. This mode was not included because it is complicated to implement in a processor that supports virtual memory. Consider a load instruction that increments its index register and then encounters a page fault in performing the memory access. If the instruction is to be reexecuted after the operating system resolves the page fault, the processor would have to "undo" the incrementing of the index register. A less-important reason for excluding it is that the 80960 architecture encourages the passing of parameters in registers instead of on the stack, meaning that the operation of pushing data onto the stack (where auto-increment/decrement is useful) is less frequent than in other architectures.

Also, the architecture was designed to allow future implementations to execute multiple instructions concurrently. In doing so, the implementation would have to detect dependencies (primarily register usage) among multiple instructions. Having load and store instructions that have the side effect of altering index registers would complicate dependency detection significantly.

ASSEMBLY-LANGUAGE NOTATION

To illustrate the semantics of instructions in later sections, it will be helpful to denote them in assembly-language format. The assembly language is summarized here; the syntax is that of Intel's assembler.

Instructions are specified by an opcode mnemonic followed by zero or more operands. In general, the operand serving as the result is the rightmost one. In REG- and COBR-format instructions, the order of the operands is source1, source2, and src/dest. Local register numbers begin with r; global register numbers begin with g.

Examples are shown below. Note that for load and store instructions, the assembler makes the decision as to whether the MEMA or MEMB format is used.

```
addi      g5,g6,g7          #  g5 + g6 -> g7
addi      1,g6,r7           #  1 + g6 -> r7
lda       0xF0,r8           #  put the hexadecimal value F0 into
                               register r8 (padded to 000000F0)
stob      g2,3(g4)          #  store low-order byte of g2 into
                               byte in memory at address 3 + g4
ld        4(r7),r13         #  load the word in memory at address
                               4 + r7 into register r13
ld        (r7),r13          #  load the word in memory whose
                               address is in r7 into register r13
ld        abc[r6*2],r13     #  load the word in memory at address
```

```
                                        abc + r6 scaled by 2 (shifted left
                                        one bit) into register r13
   ld          abc(r7)[r6*1],r13   #    load the word in memory at address
                                        abc + r6 + r7 into register r13
```

INSTRUCTION CONCURRENCY

A unique aspect of the 80960 architecture is that it permits the processor to execute instructions out-of-order or concurrently. The obvious motivation is to allow the design of very-high-performance processors.

Specifically, for any sequence of instructions i and $i+1$, instruction $i+1$ can be executed prior to, or concurrently with, i if (1) the result of $i+1$ is independent from the result of i, (2) the execution of $i+1$ does not change the input values of i, (3) the input values of $i+1$ are independent from the result of i, and (4) tracing is not enabled. "Result" means any effect or side-effect, other than a fault. Another way of saying the above is that i and $i+1$ can be executed concurrently or out-of-order "if common sense indicates so."

As an example, the architecture permits the following sequence of four instructions to execute in any order or simultaneously

```
   ld          (r4),r5
   addi        1,r6,r6
   addr        fp0,fp1,fp1
   b           loophead
```

Effective use of concurrent instruction execution requires that the processor decode multiple instructions simultaneously and that it detect dependencies among instruction operands. The first-generation processor does not attempt to do the former, and the latter is done with the register scoreboard only in the case of LOAD instructions. Other than the use of the scoreboard for LOAD instructions, the only instance of concurrent execution is branching instructions. For instance, in the following sequence, the processor will usually arrive at instruction yyy before the MULTIPLY instruction has completed.

```
             muli        r6,25,r6
             b           xxx

   xxx:      b           yyy

   yyy:
```

The major implication of concurrent execution is on fault conditions. The architecture allows certain types of faults to be *imprecise faults*, which means that when a fault such as integer overflow occurs, the architecture accurately reports the address of the instruction causing the fault, but it does not guarantee that subsequent instructions have not been executed.

The concept of imprecise faults causes no problems in implementing compilers for most programming languages, because most languages do not provide the program with the capability of resuming execution at the point of the fault. The Ada language specification, for instance, specifically allows the compiler or processor to reorder instructions.

There are, however, two problems with imprecise faults. The first is that Ada programs are permitted to define new exception handlers when they cross block boundaries. This means that an Ada compiler must not permit concurrent execution at block boundaries. The SYNCF (synchronize-faults) instruction is provides this. When executed, it will suspend instruction execution until any imprecise faults associated with any currently executing instructions occur. The second problem is that the PL/I language permits exception handlers (ON-units) to resume execution at the point of an exception (fault). To solve these problems, the architecture contains a flag called *nif* (no imprecise faults) that specifies whether all faults need to be precise. However, use of *nif* will result in slower execution (although, in the initial implementation, it does not).

INSTRUCTION SET

Most of the instructions have the REG format and are summarized in Tables 2-3 and 2-4. Most of the instructions use three register fields, where *source1* and *source2* are input values and the result is stored in the register specified by *src/dest*. In instructions that operate on operands larger than 32 bits, the operand is defined to be in contiguous registers, and the register field specifies the first (lowest-numbered) register.

The "result" column shows the value stored in the destination register(s), where A denotes the value of the *source1* operand and B denotes the value of the *source2* operand. The tables also indicate which instructions alter the value of the condition code. The number of instructions that set the condition code was held to a minimum to allow compilers to reorder instructions to exploit the condition-code scoreboard.

Table 2-3:   REG-format instructions

Mnemonic	Function	Result	CCode set?
mov	1-word move	A	
movl	2-word move	A	
movt	3-word move	A	
movq	4-word move	A	
addo	add (unsigned)	B + A	
addi	add	B + A	
subo	subtract (unsigned)	B - A	
subi	subtract	B - A	
mulo	multiply (unsigned)	B × A	
muli	multiply	B × A	
divo	divide (unsigned)	B ÷ A	
divi	divide	B ÷ A	
remo	remainder (unsigned)	B - (B ÷ A) × A	
remi	remainder	B - (B ÷ A) × A	
modi	modulo	B - (B ÷ A) × A	
shlo	shift left (unsigned)	B << A	
shli	shift left	B << A	
shro	shift right (unsigned)	B >> A	
shri	shift right	B >> A	
shrdi	dividing shift	$B \div 2^A$	
addc	add with carry	B + A + c	X
subc	subtract with carry	B - A - 1 + c	X
and	and	B & A	
notand	not B and A	~B & A	
andnot	B and not A	B & ~A	
xor	exclusive or	B ⊕ A	
or	or	B | A	
nor	nor	~(B | A)	
xnor	exclusive nor	~(B ⊕ A)	
not	not	~A	
notor	not B or A	~B | A	
ornot	B or not A	B | ~A	
nand	nand	~(B & A)	

c represents the middle condition-code bit

Table 2-4: REG-format instructions (continued)

Mnemonic	Function	Result	CCode set?
rotate	rotate B left by A mod 32	see text	
cmpo	compare (unsigned)	A ? B	X
cmpi	compare	A ? B	X
cmpinco	compare and increment (unsigned)	B + 1	X
cmpinci	compare and increment	B + 1	X
cmpdeco	compare and decrement (unsigned)	B - 1	X
cmpdeci	compare and decrement	B - 1	X
scanbyte	compare each byte		X
setbit	set a bit	B | 2Am32	
clrbit	clear a bit	B & ~2Am32	
notbit	invert a bit	B $\oplus$ 2Am32	
chkbit	check a bit		X
alterbit	alter a bit	see text	
scanbit	find most-significant 1	see text	X
spanbit	find most-significant 0	see text	X
extract	extract a field	see text	
modify	or under mask	see text	
flushreg	flush local registers		
calls	call system		
atadd	atomic add	see text	
atmod	atomic modify	see text	
modac	modify arithmetic controls	see text	X
modpc	modify process controls	see text	
modtc	modify trace controls	see text	
syncf	synchronize faults		
mark	mark (breakpoint)		
fmark	force mark		
synmov	synchronous 1-word move	see text	X
synmovl	synchronous 2-word move	see text	X
synmovq	synchronous 4-word move	see text	X
synld	synchronous 1-word store	see text	X

2Am32 represents the value 2A modulo 32

Note that many of the instructions have two forms: an ordinal (unsigned) and integer form. Ordinal arithmetic is modular arithmetic, meaning if the result exceeds the range of the destination, the higher-order bits are truncated. Integer arithmetic is signed, and the integer-overflow fault can occur (if not disabled) if the result is greater than the range of the destination. In most cases, when an overflow occurs (whether masked or not), the N least-significant bits of the result are stored in the destination (where N is the size of the destination).

The MOVE instructions copy the value of one, two, three, or four registers into the corresponding number of other registers. Multiword moves are useful when dealing with large operands and when copying parameters from global registers to local registers.

The REMAINDER instructions perform a division and return the remainder instead of the quotient. In the integer REMAINDER (REMI) instruction, the sign of the remainder is that of the dividend (B). The integer MODULO (MODI) instruction is similar, except that the sign of the remainder is that of the divisor (A).

In addition to providing the conventional signed and unsigned left and right shifts, a special signed right shift, SHRDI, is provided. It is desirable for a compiler or programmer to use a shift instruction instead of a divide instruction when dividing a number of a power of two, but conventional right shifts do not produce the correct result when the number is a negative odd number. For instance, $-1 \div 2$ should produce a result of 0, but doing this by a right shift (SHRI) yields a result of -1. The SHRDI instruction produces the correct result by conditionally adding 1 to the result if the value being shifted is negative and odd.

The next two instructions, ADDC and SUBC, are provided for unsigned and signed arithmetic on values larger than 32 bits. They operate on ordinal operands. The ADDC (add with carry) instruction adds the two source values to bit 1 of the condition code. In addition to producing a 32-bit result, it sets the condition code to $0CV$, where C is the carry output bit of the addition, and V is set if the addition would have generated an overflow had it been an integer (signed) addition. The SUBC instruction is similar.

The ALU (arithmetic/logic unit) in the processor provides all 16 possible logical operations between two values. Although many of these operations are rarely used, it was easier to provide them in the instruction set than to provide only the typical small subset. However, since 5 of the 16 operations are redundant with other instructions, only the remaining operations are provided.

The ROTATE instruction rotates the B operand to the left by the amount A modulo 32.

The six comparison instructions compare two source values and set the condition code to 100 (A<B), 010 (A=B), or 001 (A>B). The CMPINCI and CMPINCO (compare and increment) instructions also produce a result. After performing the comparison, they store the value B+1 in the destination register. These are intended for iterative loops. For instance, the following C code on the left might produce the assembly code on the right:

```
int i;
int a[10];
for (i=0;i<10;i++) {              lda      a,g8          #g8 points to a
    a[i] = i;    }               mov      0,g9          #use g9 as i
                       loop1:    st       g9,(g8)[g9*4]
                                 cmpinci  10,g9,g9
                                 bg       loop1
```

The CMPDECO and CMPDECI instructions are similar, except that they produce the result B-1. Although the CMPINCI and CMPDECI instructions perform a signed comparison and arithmetic operation, overflow is suppressed. The reason is that, if overflow could occur, these instructions could not be safely used in some languages for loops counting up or down to some variable limit (where the limit could be the maximum-valued positive or negative integer).

The SCANBYTE instruction does a byte-by-byte comparison of two source values and sets the condition code to 010 if any of the corresponding bytes are equal, or to 000 otherwise. By setting one of the source values to 0, one can use this to determine if any byte in the other operand is a 0 in one cycle. This peculiar instruction has a specific purpose: to be used in implementing the string functions in the C language. In C, the length of a string is determined implicitly by a trailing null character (00). Thus the C string functions must examine all string characters for this null value during the process of copying, comparing, of determining the length of, strings. The following code implements the C library function strcpy; it assumes the compiler allocates all strings on word boundaries. Note how the code is optimized to make use of the SCANBYTE instruction, the register scoreboard, and the condition-code scoreboard.

```
#define src_addr    g3
#define dst_addr    g2
#define word        r4
#define word_x      r5
#define mask        r6
                    proc (_strcpy)
                    ld        (src_addr), word_x      # initial read
                    lda       0xff, mask
word_loop:          addo      4, src_addr, src_addr
                    scanbyte  0, word_x               # find null?
                    mov       word_x, word
                    be        char_loop               # if yes, exit word loop
                    ld        (src_addr), word_x      # start next read
                    st        word, (dst_addr)        # store last word
                    addo      4, dst_addr, dst_addr
                    b         word_loop
```

```
char_loop:       and      mask, word, word_x
                 shro     8, word, word
                 cmpo     0, word_x              # is this byte null?
                 stob     word_x, (dst_addr)     # store the byte
                 addo     1, dst_addr, dst_addr
                 bne      char_loop              # if no, extract another
                 ret
```

A set of bit-manipulation instructions is provided. They are best illustrated by example, as shown below.

```
mov       5,g0           # register g0 = 000...000101
setbit    1,g0,g1        # register g1 = 000...000111
mov       0,g8
clrbit    g8,g1,g1       # register g1 = 000...000110
notbit    3,g1,g1        # register g1 = 000...001110
chkbit    0,g1           # condition code = 000
chkbit    1,g1           # condition code = 010
alterbit  0,g1,g2        # register g2 = 000...001111
scanbit   g2,g3          # register g3 = 3,  condition code = 010
spanbit   g2,g3          # register g3 = 31, condition code = 010
mov       0,g2
scanbit   g2,g3          # register g3 = -1, condition code = 000
```

The CHKBIT instruction moves a specified bit into the condition code (condition code = 000 or 010), and ALTERBIT moves the middle condition-code bit into a specified bit. The SCANBIT instruction searches for the most-significant set bit; it either returns the number of this bit, or if none returns a -1 value and a different condition code. The SPANBIT instruction is similar, except it searches for the most-significant clear (0) bit.

The EXTRACT instruction moves a designated bit field from one operand into another. Since it requires three input values (bit-field position and length, and the word containing the field), the *src/dest* operand is used both as a source value and the result. The architecture does not contain a corresponding insert instruction, because such an instruction would require five operands (the field to be inserted, the length and position of the field in the target, the word into which the field should be inserted, and the result). Instead, a MODIFY instruction is provided that performs a "partial insertion," which is the function

$$C = (B \ \& \ A) \ | \ (C \ \& \ {\sim}A)$$

C is the *src/dest* operand, which is the word into which a bit field is to be placed. A is a mask and B contains the field to be inserted. To perform a general insert operation, MODIFY needs to be preceded by several instructions, one which shifts the field (B) to the proper position, and one or more that construct the mask. This is illustrated by the following example, which copies bits 2-5 of register g3 into bits 2-5 of register g4.

```
extract    2,4,g3        # g3 = 000...000xxxx
shlo       28,g3,g3      # g3 = xxxx000...000
shlo       28,15,g5      # g5 = 1111000...000
modify     g5,g3,g4      # g4 = xxxxyyy...yyy
```

In this implementation of the processor, these operations are slower than the equivalent set of more-primitive operations, when the attributes of the field (position and length) are statically known (which is the case in the above example). Hence, in the current chip (but not necessarily the case in future processors), the following code sequence is considerably faster than that above.

```
shro    2,g3,g3
shlo    28,g3,g3      # g3 = xxxx000...000
shlo    4,g4,g4       # g4 = yyy...yyy0000
shro    4,g4,g4       # g4 = 0000yyy...yyy
or      g3,g4,g4      # g4 = xxxxyyy...yyy
```

The remaining instructions in Tables 2-3 and 2-4 are discussed later in this chapter.

The next group of instructions have the CTRL format, meaning that they consist of an 8-bit opcode and a 22-bit word displacement; they are listed in Table 2-5.

The BRANCH (b) instruction adds the displacement (which is signed) to the IP (instruction pointer), which results in a branch. The BAL instruction does the same, and in addition stores the address of the next instruction in a register. Since the CTRL format does not contain a register-address field, the register used is g14. This is the only instruction in the architecture that is tied to a specific register.

Table 2-5:   CTRL-format instructions

Mnemonic	Function
b	branch
bal	branch and link
bno	branch if not ordered
bg	branch if greater
be	branch if equal
bge	branch if greater or equal
bl	branch if less
bne	branch if not equal
ble	branch if less or equal
bo	branch if ordered
faultno	fault if not ordered
faultg	fault if greater
faulte	fault if equal
faultge	fault if greater or equal
faultl	fault if less
faultne	fault if not equal
faultle	fault if less or equal
faulto	fault if ordered
call	procedure call
ret	procedure return

Depending on one's point of view, there are one or eight conditional-branch instructions. The low-order three bits of the opcode are used as a mask applied to the condition code. If the result of anding the mask with the condition code is not zero, or if both the mask and condition code are zero, the branch is taken (by adding the displacement to the IP). The mnemonics were selected to describe the action of each branch in the case where the condition code is the result of a COMPARE instruction. The BO and BNO instructions (branch if ordered/unordered) are intended for use with floating-point comparisons (see Chapter 3).

Note that some instructions set the condition code to 010 (true) or 000 (false). The appropriate branches to use are BE for branching if true, and BNO (not BNE) for branching if false. The Intel assembler provides two alternative mnemonics for these situations: BT (branch if true, same as BE) and BF (branch if false, same as BNO).

The conditional-fault instructions are similar, except that they generate a fault instead of a branch. The displacement field in these instructions is unused. The rationale for these instructions is that conditional branches can cause pipeline breaks, where execution is delayed until the processor determines whether the branch will occur. For situations where program flow continues sequentially almost all of the time (e.g., performing range checks in an Ada program), the FAULT instructions can lead to better performance.*

The CALL and RETURN instructions are discussed later.

The instructions having the COBR format, listed in Table 2-6, have a format that is a cross between the REG format and the CTRL format. They consist of two register fields and an 11-bit signed word displacement.

The first set, the compare-and-branch instructions, are equivalent to a compare instruction followed by a conditional branch. Similarly, the BBS and BBC instructions are equivalent to a CHKBIT instruction followed by a conditional branch. For instance, the following sequences are equivalent:

```
cmpible    0,r8,found          cmpi      0,r8
                               ble       found

bbs        1,r8,found          chkbit    1,r8
                               be        found
```

The compare-and-branch instructions present one with a code-generation dilemma, because they have two advantages and one disadvantage when compared to the separate compare/branch instructions. One obvious advantage is that they take half the space (one word instead of two), and this can be a performance advantage in terms of the instruction cache hit ratio. Also, they allow the processor to "see" an upcoming branch operation one instruction sooner, which could be used to improve performance (although the initial implementation does not take advantage of this).

---

*But the best-laid plans often go awry. In the first-generation chips, the conditional-fault instructions *do* cause pipeline breaks, as shown in Chapter 8, and in some cases can actually be slower than untaken branches. But, the potential exists for fixing this in the next implementation.

Table 2-6: COBR-format instructions

Mnemonic	Function	CCode set?
cmpobg	compare, branch if greater	X
cmpobe	compare, branch if equal	X
cmpobge	compare, branch if greater or equal	X
cmpobl	compare, branch if less	X
cmpobne	compare, branch if not equal	X
cmpoble	compare, branch if less or equal	X
cmpibg	compare, branch if greater	X
cmpibe	compare, branch if equal	X
cmpibge	compare, branch if greater or equal	X
cmpibl	compare, branch if less	X
cmpibne	compare, branch if not equal	X
cmpible	compare, branch if less or equal	X
cmpibo	compare, branch if ordered	X
cmpibno	compare, branch if not ordered	X
bbs	check bit, branch if set	X
bbc	check bit, branch if clear	X
testno	test for not ordered	
testg	test for greater	
teste	test for equal	
testge	test for greater or equal	
testl	test for less	
testne	test for not equal	
testle	test for less or equal	
testo	test for ordered	

The disadvantage is that they do not allow one to exploit the condition-code scoreboard. A general rule-of-thumb is to use separate compare and branch instructions when one can rearrange code to use the condition-code scoreboard, and to use the combined instruction otherwise. Thus, for the following program fragment, the rightmost code will usually be faster (at least in the first-generation chips).

```
a = b + c;              addo     r5,r6,r4        cmpo    0,r7
if (d = 0) a = a + 1;   cmpobne  0,r7,x          addo    r5,r6,r4
                        addo     1,r4,r4         bne     x
                     x: ...                      addo    1,r4,r4
                                              x: ...
```

There are two instructions in the list that serve no apparently useful purpose. The CMPIBNO instruction compares two values and never branches, and the CMPIBO compares two values and always branches. Although one could possibly find an occasional use for the latter, they exist for reasons of implementation orthogonality. The ordinal forms of these two instructions do not exist because their opcodes are used for the BBS and BBC instructions.

The TEST instructions are used to test the condition code and set a boolean value in a register. The *source1* field specifies the register, and the other register field and destination are unused. For instance, the following code might be generated for the C statement on the left.

```
a = b > c;          cmpo    r5,r6
                    testg   r4      # r4 set to 0 or 1
```

Without the TEST instructions, one would have to do the above with branches, which would create pipeline breaks.

The remaining instructions have the MEM format and are listed in Table 2-7. All instructions in this category specify an effective address, which is used as a memory address.

The LOAD instructions have a one-byte, two-byte, word, two-word, three-word, and four-word form. The multiple-word loads take advantage of the burst-transfer capability of the processor's bus, which allow one to four words to be transferred from a single address. The multiple-word loads are useful for transferring large operands (e.g., floating-point values), and in implementing algorithms that transfer large amounts of successive data from memory (e.g., string algorithms).

For the one- and two-byte loads, two forms are provided: integer and ordinal. The distinction is how the value is padded when loaded into a register. For the ordinal LOADs, the value is zero-extended; for the integer LOADs, the value is sign-extended.

A corresponding set of STORE instructions is present. Again, for the one- and two-byte stores, there are two forms. The ordinal STOREs simply place the low-order one or two bytes of the specified register into the memory location. The integer STOREs do the same, except that they generate an overflow fault if the 32-bit signed value of the register to be stored cannot be represented in the memory destination (e.g., for STIB, overflow occurs if the value in the register is less than -128 or greater than 127).

The LOAD-ADDRESS instruction loads the effective address into the specified register. As mentioned earlier, this instruction has many uses, such as loading a 12-bit constant into a register or adding a 12-bit constant to a register.

Table 2-7: MEM-format instructions

Mnemonic	Function
ld	load word
ldl	load long (2 words)
ldt	load triple (3 words)
ldq	load quad (4 words)
ldob	load byte (zero extended)
ldos	load short (2 bytes, zero extended)
ldib	load byte (sign extended)
ldis	load short (2 bytes, sign extended)
st	store word
stl	store long (2 words)
stt	store triple (3 words)
stq	store quad (4 words)
stob	store byte (truncated)
stos	store short (2 bytes, truncated)
stib	store byte (overflow possible)
stis	store short (2 bytes, overflow possible)
lda	load address
bx	branch extended
balx	branch and link extended
callx	call extended

This format also includes a form of the BRANCH, BRANCH-AND-LINK, and CALL instructions. This allows one, for instance, to specify the target instruction via a full 32-bit address or as a value in a register. As an example, the following sequence selects the *ith* word from a table of branch addresses and branches to that address.

```
ld      branch_table[r8*4],r8
bx      (r8)
```

REGISTER MODEL

The set of registers visible to a program are shown in Figure 2-3. It consists of 16 global registers, 16 local registers, and instruction-pointer, process-controls, trace-controls, and arithmetic-controls registers. The first two groups represent the registers normally manipulated by programs. Of these, registers g15, r0, r1, and r2 have a predefined meaning and are not normally used by programs for data storage. This leaves 28 general-purpose registers available to programs. This large set of available registers encourages the use of global flow analysis and register allocation by optimizing compilers.

The IP register holds the address of the current instruction. Since instructions must be aligned to word boundaries, the low-order two bits of this register are forced to zero by the processor. The AC register holds the condition code and arithmetic flags; it is shown in Figure 2-1. The AC register can be accessed via the MODAC instruction. The PC register holds information about the program state; it is discussed later in Figure 2-5. The TC register holds information associated with program tracing; it is discussed later in Figure 2-9. The PC and TC registers can be accessed with the MODPC and MODTC instructions.

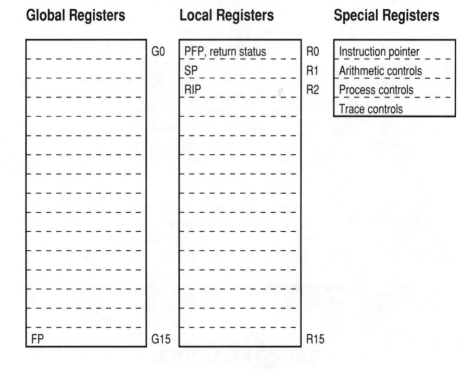

Figure 2-3:   The register set

There is one usage restriction on the general-purpose (global and local) registers. Two-word operands must be contained in even/odd register pairs, and are denoted by specifying the number of the lowest (even) register. Three- and four-word operands must be contained in consecutive registers, the first of which must have a number which is a multiple of four. For instance, the destination of a LDQ (load-quad) instruction must be specified as g0, g4, g8, g12, r0, r4, r8, or r12. The processor does not check to see if this is the case (although the assembler does); failing to adhere to these register-alignment rules causes an undefined set of registers to be accessed.

The rationale for these register-alignment rules will be obvious to anyone who has implemented a high-speed processor. The rules allow the processor to form the addresses of successive registers for multiword operations by simply manipulating the low-order one or two bits of the register number. Allowing arbitrary alignment of multiword register operands would require the processor to have adders for register numbers on critical circuit paths.

PROCEDURE CALL/RETURN

The CALL and RETURN instructions allocate and deallocate stack frames in a manner similar to other architectures. However, because of the way local registers are used, most call and return operations perform no memory references.

There are three aspects of the local registers that are pertinent to procedure calls and returns:

1. The processor has multiple sets of local registers. The number of sets is not specified by the architecture; the first-generation chips have four sets. When a call instruction is executed, the processor allocates a set to the called procedure.

2. Saving and restoring of local registers (i.e., when the processor does not have an available set) is done transparently to the program. When a register set needs to be saved, it is saved in the first 16 words of the stack frame associated with the register set.

3. Subroutine (stack frame) linkage information is kept in the local registers.

There are four registers associated with procedure call/return: global register g15 and local registers r0-r2. Register g15 contains the address of the current stack frame. r1 points to the top of the stack. r2 contains the instruction pointer to be used when returning to the frame. r0 contains a pointer to the previous frame in bits 4-31, a *prereturn trace* indicator in bit 3, and the *return status* in bits 0-2.*

When a call operation is performed, the processor stores the address of the next instruction in register RIP (r1), allocates a new set of registers, stores the address of the previous frame, with some status information, in PFP (r0), stores the address of the new stack frame in FP (g15), and stores the address of the current frame plus 64 in SP (r1).

After a call operation, the three low-order bits of r0 contain status information used by the return instruction. The purpose of this status information is to allow the architecture to have only a single return instruction (e.g., so that interrupt routines do not have to exit with a special interrupt-return instruction, which would require special compiler support and not allow these routines to be called by other routines). The state of the status bits after a call operation is one of the following:

000	- local call, or supervisor call from supervisor mode
001	- fault call (implicit call of a fault routine)
010	- supervisor call from user mode, tracing was disabled
011	- supervisor call from user mode, tracing was enabled
111	- interrupt call while in the executing state

One more state bit exists in r0, the *P*, or prereturn-trace, flag. This is associated with detection of the prereturn trace event when executing a RETURN instruction. Call operations set this bit to 0 unless a call trace event occurs, in which case the bit is set to 1.

The remaining part of the call operation that needs to be discussed is how the processor allocates stack frames in memory. Rather than allocating the new frame at the end of the current frame, the new frame is allocated on a specific boundary beyond the end of the current frame. The architecture specifies that this boundary must be an implementation-defined multiple of 16 bytes (i.e., at least on a 16-byte boundary); the first-generation chips locate frames on a 64-byte boundary. In other words, stack frames must be at least 16 words in size (in case the local registers associated with the frame need to be saved). This is the reason that the call operation initializes register r1 (SP, stack pointer) to the frame pointer plus 64. If the procedure does not need a frame larger than 64 bytes, the next frame will begin just after the current frame. If the procedure needs more frame space (and thus changes SP after being called), the next frame will begin on the next 64-byte boundary.

There are two reasons for the alignment of stack frames. First, the external bus is optimized for multiword transfers that start on 16-byte boundaries; thus the saving and restoring of local registers can take advantage of this. Second, since the

---

*Because of these semantics, in the Intel assembler registers g15, r0, r1, and r2 are named fp, pfp, sp, and rip, respectively.

low-order four bits of r0 are used for status flags, the previous frame pointer is restricted to pointing to a 16-byte boundary.

Given this background, the function of the call instructions (CALL and CALLX) can now be expressed as the following algorithm.

```
wait for any uncompleted instructions to finish;
temp = (SP + 63) & 0xFFFFFFC0; /* round up if necessary */
RIP  = IP;                /* RIP is r2 */
if (register_set_available) allocate local register set;
    else {save a register set in memory at its FP;
          allocate local register set;}
IP  = target_address;
PFP = FP;                /* store g15 in r0, status bits are 0 */
FP  = temp;
SP  = temp + 64;
```

The algorithm of the RETURN instruction is shown below. Some of the functions pertain to returning from other types of call operations, which are discussed later in this chapter.

```
wait for any uncompleted instructions to finish;
switch (frame_status) {

        case 0x0:        local_return();   /* Local return */
                         break;

        case 0x1:            /* Fault handler return */
                         temp  = memory(FP-16);
                         temp1 = memory(FP-12);
                         local_return();
                         arithmetic_controls = temp1;
                         if (process_controls.emode == supervisor)
                             process_controls = temp;
                         break;

        case 0x2:            /* Supervisor return, disable trace */
                         if (process_controls.emode == supervisor)
                             {process_controls.T = 0;
                              process_controls.emode = user;}
                                 local_return();
                         break;
```

```
                case 0x3:                /* Supervisor return, enable trace */
                             if (process_controls.emode == supervisor)
                                 {process_controls.T = 1;
                                  process_controls.emode = user;}
                                      local_return();
                             break;

                case 0x8:                /* Interrupt return to executing state */
                             temp  = memory(FP-16);
                             temp1 = memory(FP-12);
                             local_return();
                             arithmetic_controls = temp1;
                             if (process_controls.emode == supervisor)
                                 {process_controls = temp;
                                  check_pending_interrupts();}
                                      /* if continue here, no interrupt to handle */
                             break; }


        void local_return()
        {              FP = PFP;
                       free current local register set;
                       if (register_set(FP) not present) retrieve from memory(FP);
                       IP = RIP;

        }
```

The actions associated with returns from fault and interrupt handlers are discussed in later sections.

Note that the type of return performed depends on the frame-status information
in register r0, which is a register accessible to the program. Thus the return operations that perform more than a simple local return do so only if the program is in
supervisor mode.

Although the allocation of registers is a decision made by software, it is the
intent of the architecture that local variables be allocated directly in local registers
and parameters be passed in global registers (as opposed to pushing them on the
stack). Since procedures tend to have a small number of local variables, this means
that most procedures will not require allocation of local variables on the stack in
memory. For instance, a study of stack-frame sizes for UNIX commands[3] shows
that about 90% of the procedures would fit within the local register space. In a
prototype port of the UNIX System V kernel to the 80960 protected architecture,
98.8% of the procedure calls passed fewer than six words of arguments, and 97.1%
of the functions required no stack space beyond that of the local registers.

An example of a convention for register allocation and parameter passing (using the current Intel compilers as the example) is given in Chapter 8.

Only the current local-register set is accessible to software, but there are situations where one needs access to the contents of the other register sets (e.g., if one makes an uplevel reference to a local variable of a calling procedure). The FLUSHREG instruction writes all local-register sets, except for the current one, to their associated stack frames in memory, and marks them as purged.

Unlike most other architectures, the stack grows "upward." There are two reasons for this. The first reason is the burst characteristics of the external bus, which allows multiple successive words to be transferred by specifying the address of the word at the lowest address. If the stack grew downward, the processor would have to perform an extra computation (FP-64) in order to determine the starting address for a transfer of a local-register set to/from memory. The second reason pertains to the protected architecture (Chapter 4). Like most other architectures, page tables can be variable in size and expand "upward." Thus, if a program runs out of stack space, the operating system can extend the stack space by adding an entry to the appropriate page table. The VAX architecture solves this problem by providing a special page table that can be expanded downward.[4]

SUPERVISOR CALL/RETURN

The architecture provides a two-level privilege model (user and supervisor states) and a mechanism that can be used to call operating-system procedures and enter supervisor state. The core architecture, however, does not provide any mechanism for memory protection, and thus the use of user/supervisor privileges is of limited use. The protected architecture (Chapter 4) does provide memory protection and associates it with user/supervisor privilege.

The CALLS (call-system) instruction specifies a value instead of an address. This value is used as an index to an entry in the system procedure table. This table, whose address is established at initialization, is shown in Figure 2-4.

The table contains a stack pointer, a trace (T) flag, and 260 procedure-address entries.[*] If the two low-order bits of a procedure-address entry are 10, the procedure is a supervisor procedure; if the bits are 00, the procedure is a local procedure.

---

[*]260 is an "odd" number (e.g., versus 256). The rationale for 260 is how data structures are mapped in an extension of the architecture, and the need for the core architecture to be upward compatible to it.

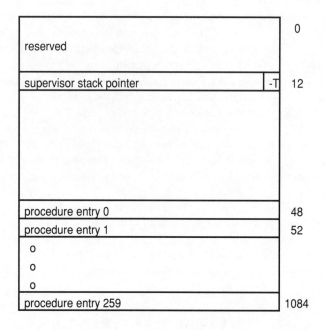

Figure 2-4:   System procedure table

The CALLS instruction is similar to the call instruction, except that it performs the following additional actions (and it gets the procedure address from the table). If the entry is for a local procedure, or the program is already in supervisor mode, the operation is the same as that for a CALL instruction. Otherwise, a switch is made to a new stack (the new frame begins at the stack address specified in the table), the execution mode is set to supervisor, the frame status in register r0 of the new frame is set to 01T, where T is the value of bit 0 of the process-controls register, and the T flag from the table is stored in bit 0 of the process-controls register.

The presence of the T flag in the table allows one to have tracing enabled in an application program and to disable tracing when it invokes the operating system. The saving of the caller's trace flag in the frame status allows the return instruction to restore it when returning back to the application program.

SYNCHRONIZATION

In multiple-processor systems, a mechanism is needed to allow programs to manipulate shared data structures in an indivisible manner, meaning that when such an operation is underway, another processor cannot perform the same operation. The architecture provides two instructions for this purpose; these instructions can be used by programs to implement higher-level synchronization mechanisms, such as locks and semaphores. The ATADD (atomic-add) instruction performs the following operation:

```
temp = atomic_read(A);
atomic_write(A) = temp + B;
C = temp;
```

As before, A, B, and C represent, respectively, the *source1*, *source/2*, and *src/dest* operands.

The ATADD instruction adds a value to a word in memory and returns the original value of the word. The atomic_read operation waits until the LOCK line on the external bus is not asserted, and then asserts the LOCK line and performs the read. The atomic_write operation performs a write operation and deasserts the LOCK line. This ensures that another processor cannot perform an atomic_read operation between the read and write of the word in memory.

The ATMOD (atomic-modify) instruction performs the following operation:

```
temp = atomic_read(A);
atomic_write(A) = (C & B) | (temp & B);
C = temp;
```

This logical expression, *or under mask*, exists in several other instructions. It allows the ATMOD instruction to set or clear one or more bits in a memory location indivisibly. The following code uses ATMOD to implement a primitive spin lock.

```
        lda      statusword,r4
        lda      1,r5
spin:   atmod    r4,1,r5      # set bit 0 in statusword, r5 = original value
        and      1,r5,r5      # clear all but bit 0
        cmpobe   1,r5,spin    # try again if bit 0 was already set
```

PROCESS CONTROLS

Another control register in the processor (similar to the arithmetic-controls register) is the process controls (PC), shown in Figure 2-5.

The *internal-state* field is not described as part of the architecture; it is an implementation-dependent field. The *priority* field is associated with the priority-interrupt mechanism in the next section; it defines the current priority of the processor, 0 being lowest and 31 being highest. The bits designated as 0 are associated with features in higher levels of the architecture, and must be set to 0 by software.

The *state* flag specifies whether the processor is in the interrupted state (1) or not (0). The *execution-mode* flag specifies whether the processor is in supervisor (1) or user (0) mode.

The remaining two flags are associated with the program-trace mechanism. The *trace-enable* flag is set if tracing is enabled. The *trace-fault-pending* flag is set if a trace fault should occur prior to executing the next instruction; this flag is used by the processor and should not be set by software.

```
31                                                                              0
┌──────────────────────┬──────────┬──┬─┬──┬─┬──┬───┬─┬─┬──────┬──┬─┐
│   internal state     │ priority │--│0│st│0│--│tfp│0│0│  ---  │em│t│
└──────────────────────┴──────────┴──┴─┴──┴─┴──┴───┴─┴─┴──────┴──┴─┘
```

BITS

internal state	21-31	internal state - implementation dependent
priority	16-20	priority (0-31)
0	14	must be zero
st	13	state (0 = executing, 1 = interrupted)
0	12	must be zero
tfp	10	trace fault pending
0	9	must be zero
0	8	must be zero
em	1	execution mode (0 = user, 1 = supervisor)
t	0	trace enable

Figure 2-5:   Process controls

An instruction, MODPC, is provided to allow software to read and modify the process controls. The operation performed by this instruction is shown below.

```
if (B != 0) {
    if (process_controls.emode = supervisor) {raise type_mismatch fault}
    temp = process_controls;
    process_controls = (B & C) | (process_controls & ~B);
    C = temp;
    if (temp.priority > process_controls.priority) {check_pending_interrupts}
        /* if continue here, no interrupt to handle */
}
else C = process_controls;
```

Like the other types of modify instructions, this instruction allows one to alter the fields in the process controls selectively. However, it performs two additional functions. First, it requires the program to be in supervisor state if the mask is not zero (i.e., if one is going to change the process controls); a user-state program is permitted to read the process controls. Second, if the priority in the process controls is lowered, the processor checks for pending interrupts (to find out if there is now a higher-priority interrupt that needs to be handled).

Another instruction in this class exists to read and/or modify the arithmetic controls. The MODAC instruction performs the following function.

```
temp = arithmetic_controls;
arithmetic_controls = (B & A) | (arithmetic_controls & ~A);
C = temp;
```

INTERRUPTS

Unlike most other microprocessor architectures, the architecture contains a priority-based interrupt mechanism. An interrupt is an event, usually generated externally, that has an associated priority. If the interrupt priority is 31, or is higher than the current priority of the processor, an interrupt handler is invoked. Otherwise, the interrupt is remembered for later processing by setting a bit in a bit vector.

Each interrupt is associated with an 8-bit vector number, which associates the interrupt with an entry in the interrupt table. The interrupt table is a data structure whose address is defined at the time of processor initialization. Its structure is illustrated in Figure 2-6.

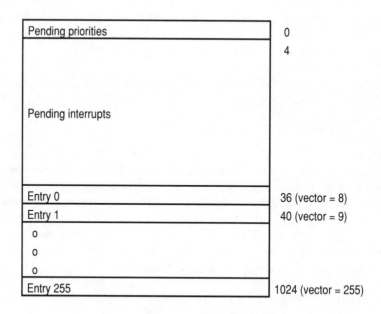

Figure 2-6: Interrupt Table

The priority of an interrupt is its vector number divided by eight. Thus, there are eight interrupts that can be defined at a specific priority level, and 32 priority levels, for a total of 256 interrupts. However, priority-0 interrupts serve no purpose, since they could never interrupt the processor (since the lowest processor priority is 0). Thus, this group is excluded, meaning that there are 248 interrupts that can be defined in the interrupt table. Each is described by a one-word entry, which contains the address of the interrupt-handler procedure associated with that interrupt.

When an interrupt handler is to be called, the following is done:

1. The value of the process controls is saved.

2. In the process controls,the state is set to interrupted, the execution mode is set to supervisor, the trace flags are cleared, and the priority is set to that of the interrupt.

3. A call operation is done to the procedure specified in the interrupt table. If the processor was in the interrupted state prior to the interrupt, the value of SP used in the call operation is SP+16; otherwise the value of SP used is ISP+16 (ISP is the interrupt stack pointer). The return status in the new r0

register is set to 111. Finally, prior to starting execution of the interrupt handler, the saved process-controls word is stored on the stack at address FP-16, and the arithmetic-controls word is stored at address FP-12.

The standard RETURN instruction is used to return from an interrupt handler. In addition to performing the normal return operations, it restores the process and arithmetic controls from the stack and checks the interrupt table for pending interrupts (the subject of the next section).

Pending Interrupts

The decision to implement a priority-based interrupt mechanism raises the issue of what to do with interrupts of lower priority. Because of the protocols involved with the signaling of interrupts (e.g., from an external interrupt controller), it is not feasible to leave them hanging at their source. Thus the processor's interrupt mechanism includes the means to remember lower-priority pending interrupts.

The interrupt table contains two bit vectors:

- one of 32 bits in length in word 0 named *pending priorities*
- one of 256 bits in length in words 1-8 named *pending interrupts.*

An interrupt of vector number i is considered pending when bit i is set in *pending interrupts*, and bit $i/32$ is set in *pending priorities*.

An interrupt is posted when the interrupt occurs and its priority is not such that the processor should immediately handle it. Posting is done by the processor via the following algorithm.

```
temp  = atomic_read(pending_priorities);  /* Synchronize */
temp1 = memory(pending_interrupts(i/8));  /* Read a byte */
temp[i/8] = 1;
temp1[i mod 8] = 1;
memory(pending_interrupts(i/8)) = temp1;
atomic_write(pending_priorities) = temp;
```

The algorithm is shown to highlight the fact that the processor sets the bits in an indivisible manner. This is done such that multiple processors can share a common interrupt table and thus share the load of interrupt handling.

Occasionally one encounters a need to have a program post an interrupt. Since there is no way for an 80960 program to perform this algorithm in software (because there is no way to have an atomic read/write operation span other instructions), a program should post interrupts by sending an interrupt IAC message.

However, in a single-processor system, it is possible to have software set the pending bits in the interrupt directly (without using an indivisible operation) if the program can guarantee that an intervening interrupt cannot occur.

The processor tests the interrupt table for pending interrupts at two principal times: during a return from an interrupt handler and at the end of a MODPC instruction that has lowered the processor priority. To test for a pending interrupt, the processor locates the most-significant set bit in *pending priorities*. If its number is higher than the current processor priority, a pending interrupt will be handled at this point. The appropriate byte in *pending interrupts* is read and the most-significant set bit in it represents the interrupt to be handled. (If it is not obvious by now, it is worthwhile to point out that *pending priorities* is redundant information with *pending interrupts*, but exists as an optimization.)

The obtaining of a pending interrupt is done in a synchronized way to permit multiple processors to post and obtain pending interrupts in the same interrupt table. Also, when a pending interrupt is obtained from the table, its *pending interrupt* bit is cleared, and the associated *pending priorities* bit is also cleared if there are no more pending interrupts at that priority.

FAULTS

In addition to interrupts, the architecture defines another type of event, *faults*, and mechanisms for handling them. Unlike interrupts, faults usually occur synchronously with program execution and represent exceptional conditions or errors in the program.

Rather than treat all fault types separately, the set of faults defined by the architecture are grouped into *fault types* and *fault subtypes*. That is, a specific fault is defined by being a particular fault type and subtype. For instance, in the fault type *trace*, there are seven subtypes, or "types" of trace faults.

Fault types and subtypes are denoted by an encoding in two 8-bit fields. The set of faults and their encodings are shown in Table 2-8. Where multiple faults can occur concurrently and this information is to be conveyed to the program, the fault subtype is encoded as a specific set bit in the 8-bit subtype field; in all other cases, the subtype is encoded as a number in the 8-bit subtype field.

Similar to the way interrupt handlers are specified, fault handlers are specified in a fault table, a data structure whose address is defined at the time of processor initialization. The fault table is illustrated in Figure 2-7. Note that the fault table contains entries for each fault type, but not subtype. The intention, for instance, is that there is a single fault handler for trace faults, and the fault subtype provides additional information about the fault.

Table 2-8: Fault types and subtypes

Type Encoding	Subtype Encoding	Description
1		Trace fault
1	xxxxxx1x	Instruction trace
1	xxxxx1xx	Branch trace
1	xxxx1xxx	Call trace
1	xxx1xxxx	Return trace
1	xx1xxxxx	Prereturn trace
1	x1xxxxxx	Supervisor trace
1	1xxxxxxx	Breakpoint trace
2		Operation fault
2	1	Invalid opcode
2	2	Unimplemented function
2	4	Invalid operand value
3		Arithmetic fault
3	1	Integer overflow
3	2	Zero divide
5		Constraint fault
5	1	Constraint range
5	2	Constraint privileged
7		Protection fault
7	2	Length
8		Machine fault
8	1	Bad access
10		Type fault
10	1	Type mismatch

Note that a fault-table entry can specify two ways of invoking the fault handler, basically via a CALL or CALLS operation. The first (local procedure) calls the fault handler using the current stack. The second (system procedure) calls the fault handler via the system procedure table. The latter would typically be done for fault handlers that are considered part of the operating system (e.g., require supervisor state). The former might be done for arithmetic faults. The "magic value" in the second word of the system-procedure entry is for compatibility with higher levels of the architecture.

When a fault is detected, the entry corresponding to the fault type is read from the fault table, and the following is done

- The value of the process controls and arithmetic controls is saved.

- A implicit CALL or CALLS operation is done to the procedure specified in the fault table. During the call, the value of the stack pointer used is adjusted upward by 16 bytes to allow space for a fault record. The return status in the new r0 register is set to 001. A fault record is copied to the stack, beginning at address FP-16.

The fault record placed on the stack is shown in Figure 2-8. The process-controls and arithmetic-controls fields contain the saved values of these registers at the time of the fault. The *ftype* and *fsubtype* fields define the fault type and subtype. Except for a few circumstances, the last field contains the address of the instruction that caused the fault.

reserved	0
Trace fault entry	8
Operation fault entry	16
Arithmetic fault entry	24
reserved	32
Constraint fault entry	40
reserved	48
Protection fault entry	56
Machine fault entry	64
reserved	72
Type fault entry	80

Local Procedure Fault-Table Entry

Procedure address	00
reserved	

System Procedure Fault-Table Entry

Procedure number	10
Must be 0000027F	

Figure 2-7: Fault table

The standard RETURN instruction is used to return from a fault handler. If the return status is 001, the arithmetic-controls register is loaded from the word at address FP-12, and, if the execution mode is supervisor, the process-controls register is loaded from the word at FP-16.

Process Controls				0
Arithmetic Controls				4
	FType		FSubtype	8
Address of Faulting Instruction				12

Figure 2-8:   Fault record

Program State after a Fault

When a fault occurs, several key questions must be answered: What changes to the program state did the faulting instruction make? What IP was saved as the result of calling the fault handler? What is the state of the instructions in the "vicinity" of the faulting instruction?

The answer to the first question is straightforward. The faulting instruction causes no state changes, meaning that the instruction can be safely reexecuted, except for the integer-overflow fault, where the destination is modified, usually with the truncated result, and the machine fault, where the effect of the instruction being executed at the time of the fault is unpredictable.

The answers to the remaining two questions depend on the type of fault and whether *nif* (no imprecise faults) is set in the arithmetic controls. Faults are grouped into three categories: *precise*, *imprecise*, and *asynchronous*. The only asynchronous fault is the machine fault. For this fault, the saved IP (as a result of the call to the fault handler) and the address of the faulting instruction in the fault record are undefined.

Precise faults are those that need to be recoverable by software. The faults in this category are trace and protection. When a precise fault occurs, the processor is prohibited from executing instructions beyond the faulting instruction if and only if those instructions will fault.* The fault record correctly identifies the faulting instruction, and the saved IP correctly identifies the point to which the fault handler might return, but the saved IP need not point to the instruction immediately after the faulting instruction (more on this later).

The remaining faults are in the imprecise category. For these faults, the address of the faulting instruction is correct, but the saved IP is unpredictable, as is the state

---

*That is, the architecture accommodates a prescient implementation. In practical terms, the implementation can execute ahead of an instruction that could result in a precise fault if it knows that the subsequent instructions are of a type that cannot fault (not all instructions can fault) or if it is able to undo the effects of the subsequent instructions in the event of a fault.

of any instructions in the "vicinity" of the faulting instruction. Furthermore, imprecise faults can occur out of order. For instance, in the following sequence, assume that an integer overflow fault will occur in both instructions. The architecture does not require the fault in the MULI instruction to occur before the fault in the ADDI instruction.


```
muli      r8,10,r8
addi      g0,1,g0
```


As is the case with other aspects of the architecture, imprecise faults are defined with future implementations in mind. To achieve significantly higher levels of performance, we anticipated that future implementations would decode and execute multiple instructions concurrently (in situations where the candidate instructions are found to be independent of one another). Requiring all faults to be precise would inhibit this in one of two ways:

1. If the test for independence includes the possibility of faults (e.g., the instructions above would fail the test because both can overflow), the probability of seeing successive independent instructions would be low (and thus the ability to execute instructions concurrently would not yield a significant performance increase).

2. Or, an alternative approach would be to execute instructions concurrently that were independent except for the possibility of faults, and include logic to undo the effects of an instruction if a concurrently executed instruction that was "earlier" in the instruction stream faulted. The logic to do this is not trivial and could add gate delays on critical speed paths in the processor. This would limit the cycle speed and thus would limit performance.

Surprisingly, perhaps, the existence of imprecise faults is not a problem for the prevalent programming languages, because such languages as Fortran and C do not provide a mechanism for recovering from faults. Ada provides for exception handlers, but does not permit exception-handling code to return to the point of the fault. In fact, the Ada specification specifically points out that the compiler/processor need not report exceptions in the order that they might appear to occur in the Ada source program. The PL/I language, however, is a problem; ON-UNITS (fault handlers) in PL/I programs are permitted to resume execution at the point of the fault. This would represent a problem if imprecise faults can occur, since the result could be that an instruction is incorrectly executed twice (e.g., in the example above, if the MULI instruction overflowed and an ON-UNIT corrected the error and resumed execution at the MULI instruction, the ADDI instruction might be executed twice, leading to g0 containing an incorrect value). To handle this, one could set the *nif* flag in the arithmetic controls, such that imprecise faults do not occur. However, in an implementation with concurrent execution, this would likely cause the processor to serialize all instruction execution.

Despite what was said previously, Ada requires that exceptions occur in order when crossing block or procedure boundaries (because different blocks or procedures can define different exception handlers). For instance, in the example above, if the MULI and ADDI instructions are in different blocks, the fault that is reported must be that in the MULI instruction (if the multiplication overflows). One could guarantee this by setting *nif*, but this could impact the performance of the overall program. A better alternative is use of the SYNCF (synchronize faults) instruction. This instruction causes the processor to wait for any faults to be generated by uncompleted instructions. Thus, the Ada compiler could generate a SYNCF instruction at the entry and exit of all blocks (actually, not at all block boundaries, but only the typically small subset of boundaries where a new exception handler is defined).

When a precise fault occurs, a second IP (in addition to the IP of the faulting instruction in the fault record) is saved in register r2 of the current frame. Depending on the type of the fault, the saved IP is defined to point to the faulting instruction, the "next" instruction, or be undefined. This saved IP is intended to be the resumption point, if the fault handler decides to return to resume execution. However, when a fault defines the saved IP as pointing to the next instruction, the architecture does not require it to point to literally the next instruction; it only need point to an instruction at which execution can be safely resumed if the fault handler decides to return to the faulting program.

The intent of this is to allow implementations to pipeline unconditional branch instructions without regard to faults. For instance, in the following sequence, assume the MULI instruction causes an overflow fault and that *nif* is set (no imprecise faults). The architecture allows the saved IP to point to either BRANCH instruction, or to the ADDI instruction, since all of these cases provide for correct resumption of the program.

```
          muli      r8,10,r8
          b         aaa
          ...
aaa:      b         bbb
          ...
bbb:      addi      g0,1,g0
```

The only out-of-order execution done by the initial chips is for BRANCH instructions (the example above, where it is likely that the saved IP would point to bbb if the instruction sequence was contained in the instruction cache). The initial chips ignore *nif* (all faults are treated as if *nif*=1), and SYNCF acts as a no-op (no operation) instruction.

TRACING

The 80960 architecture provides a mechanism for the tracing of program execution. Rather than just contain an instruction-stepping facility as is typical in most other architectures, the 80960 architecture provides for tracing on a number of distinct events. Tracing is controlled by the trace-controls register, depicted in Figure 2-9.

31   0

imp dep	trace flags	imp dep	trace modes

	BITS	
imp dep	24-31	implementation dependent
trace flags		
kte	23	breakpoint trace event
ste	22	supervisor trace event
pte	21	prereturn trace event
rte	20	return trace event
cte	19	call trace event
bte	18	branch trace event
ite	17	instruction trace event
imp dep	8-15	implementation dependent
trace modes		
ktm	7	breakpoint trace mode
stm	6	supervisor trace mode
ptm	5	prereturn trace mode
rtm	4	return trace mode
ctm	3	call trace mode
btm	2	branch trace mode
itm	1	instruction trace mode

Figure 2-9: Trace Controls

The implementation-dependent fields in the trace controls are not defined by the architecture; they control functions in a proprietary version of the chips used in

Intel's in-circuit emulator product. The *trace-modes* field defines seven distinct events that can be selectively enabled. If a specific mode bit is set, and the *t* (trace enable) flag is set in the process controls, an occurrence of the associated event results in the corresponding trace flag being set, and causes a trace fault to be generated. The reason for the and'ing of the mode bit and the *t* bit is that the *t* bit is altered by certain types of call and return operations, allowing, for instance, an application program to be traced, but the tracing is automatically disabled when the program invokes the supervisor.*

Instruction tracing results in a trace fault after the execution of every instruction. Branch tracing results in a trace fault after the execution of a branch instruction that branches (i.e., a conditional branch that does not branch does not generate a fault). This definition of branch tracing makes it useful for the tracing of control flow through a program.

Call tracing results in a trace fault after the execution of a call or branch-and-link instruction, and after the occurrence of an implicit call (e.g., as a result of a fault). Branch-and-link was included in this category (rather than in the branch-trace category) because a compiler might substitute a BAL instruction for a CALL instruction. When a call-trace event occurs, the processor sets the prereturn-trace indicator (bit 3) in register r0.

Return tracing results in a trace fault after the execution of a RETURN instruction. Prereturn tracing results in a trace fault *prior* to the execution of a return instruction, if the prereturn indicator is set in r0. A prereturn trace event also clears the prereturn indicator. The prereturn facility was put in the architecture at the request of engineers developing software debugging packages. The rationale is that when one is debugging a certain procedure in a program, there are things the debug software may wish to see upon exit of the procedure (e.g., the final state of the procedure's registers).

Supervisor tracing results in a trace fault after the execution of a CALLS instruction when the procedure being called is a supervisor procedure, and after the execution of a return instruction when the return status field in r0 is 010 or 011 (i.e., a return from supervisor mode to user mode).

Finally, breakpoint tracing results in a trace fault in three situations: (1) after executing a MARK instruction, (2) after executing an FMARK (force mark) instruction, even if the breakpoint trace mode is not set, and (3) an implementation-dependent situation. In the first-generation chips, the latter is associated with a match using any of two on-chip instruction-address breakpoint registers. The breakpoint registers are discussed later in this chapter.

The MODTC instruction can be used to read the trace-controls register, or to store information in it under control of a mask. The MODTC instruction functions the

---

*However, in the first-generation implementation, if one has one or more modes set while the *t* flag is clear, the processor slows down considerably (although not nearly as much as if a trace fault were to occur). Therefore, for full-speed operation, one should not set any trace-mode flags, even if the trace-enable flag is clear.

same way as the MODAC instruction, except the processor ignores any attempt by MODTC to alter the implementation-dependent fields in the trace controls.

Some instructions can alter the information that controls tracing. Therefore the relationship between these instructions and tracing needs further definition. Certain call and return operations can alter the t flag in the process controls (i.e., calls into supervisor mode and returns from supervisor mode). In these cases, the value of the t flag used to determine whether trace events are detected in these instructions is the value of t at the end of the instruction. For instance, if return or instruction tracing is enabled in the trace controls, a return that changes t from 0 to 1 would lead to a trace fault; a return that changes t from 1 to 0 would not. Also, the MODPC instruction can change t in the process controls and the MODTC instruction can change the trace modes in the trace controls. When this happens, the architecture definition states that the effect of the change does not need to be recognized by the processor until up to four succeeding nonbranching instructions are executed. This avoids logic complexity in processors that are highly pipelined, or in processors that execute multiple instructions concurrently.

In addition to the t flag, the process-controls register contains a flag called *tfp* (trace fault pending). This flag allows the processor to handle correctly the situation of a trace event and an interrupt occurring simultaneously. When a trace event is detected in an instruction, the only action the processor takes at this point is to set the *tfp* flag (in addition to setting the proper event flag). Before executing an instruction, the processor generates a trace fault if both t and *tfp* are set. Thus, if an interrupt occurs during an instruction in which a trace event occurs, the interrupt is handled and the trace fault occurs when the interrupt handler returns.

The prereturn trace is special in that it can cause two trace faults to occur between instructions. For instance, if both instruction tracing and prereturn tracing are enabled, we finish executing an ADDI instruction, and the next instruction is a RETURN instruction, the instruction-trace event is indicated, the trace-fault handler is called, and when the fault handler returns, the prereturn trace event occurs (calling the trace-fault handler again).

A trace-fault handler can examine the trace controls (via the MODTC instruction) to determine the event that has occurred, and can get the address of the faulting instruction from the fault record. Because of the definition of the trace events, it is possible for multiple events to occur simultaneously (e.g., if instruction, call, and supervisor trace modes are enabled, a CALLS instruction could generate all three events). However, in these cases, the architecture definition does not require the processor to mark all of the simultaneous events. Instead, if supervisor trace occurs in combination with any other event, only the supervisor event need be marked. If instruction trace occurs in combination with any other event, the instruction event need not be marked.

IMPLEMENTATION DEPENDENCIES

The 80960 architecture was defined to allow processors to have user-visible, implementation-specific, characteristics. This was done because specific implementations may need to define program-visible facilities for things like cache control, bus characteristics, multiprocessing, special-function registers, and on-chip peripherals. These facilities are specified separately to serve as a message to software engineers that these facilities might not exist, or might take on a different form, in other implementations.

In the context of the core architecture, the initial 80960KA defines the following implementation-dependent facilities:

- Processor initialization.

- Inter-agent communication messages, which are used for signaling among multiple processors, and for performing certain control operations in the processor.

- An on-chip interrupt controller.

- A set of synchronizing load and store instructions.

- On-chip breakpoint registers.

These are defined in the remaining sections of this chapter.

INITIALIZATION

Processor initialization, which is triggered by the external RESET pin, is the mechanism by which the processor begins instruction execution. In addition, the initialization mechanism embodies two other important considerations. The first is two stages of self-test function, which detect failures in the processor chip and failures in the basic system operation. The second consideration is a specification of whether the processor is an "initialization" processor or not. This is indicated by an external pin and can be used in multiprocessor systems to specify which processors should actually begin instruction execution, and which should simply initialize to the stopped state.

During the initialization stage, the processor reads eight words beginning at physical address 0, and uses some of this information to locate two other blocks of information, as shown in Figure 2-10. The block of information at the lower right side is called the *processor control block* (PRCB); one of its purposes is to contain the starting addresses of the interrupt table, interrupt stack, and fault table. The main purpose of the block at the upper right side is to contain the starting address of the system procedure table. The word at location 12 in physical memory contains the address of the first instruction to be executed.

The actual steps performed by the processor are

1. Assert the FAILURE output pin and perform the internal self-test. If the test passes, deassert FAILURE and continue with the step below; otherwise enter the stopped state.

2. Clear the trace controls, disable the breakpoint registers, clear the process controls, and then set the *em* flag in the process controls (supervisor mode). If the processor is an initialization processor, continue with the step below; otherwise enter the stopped state.

3. Read eight words from memory, beginning at location 0. Clear the condition code, sum these eight words with the ADDC (add-with-carry) operation, and then add FFFFFFFF to the sum (again with ADDC). If the sum is 0, continue with the step below; otherwise assert the FAILURE pin and enter the stopped state.

4. Use words 0 and 1 as the pointers to the initial data structures, and set the IP to the value of word 3. In the process controls, set the priority to 31 and the state to interrupted. Store the interrupt stack pointer in FP (g15), and begin execution.

The internal self-test is intended to give the user some assurance that the processor is operational. The test performs such functions as calculating a checksum on the internal ROM, writing and reading test patterns through the instruction cache and all registers, and checking the major internal data buses and ALU. The second stage of the self-test is the checksum test on the first eight words of memory. This simple test is effective in finding additional problems in the processor (e.g., defective bus driver, bonding wire, or pin), as well as finding failures in the surrounding system (e.g., short or open in the bus, missing ROMs, failures in the external bus-support logic).

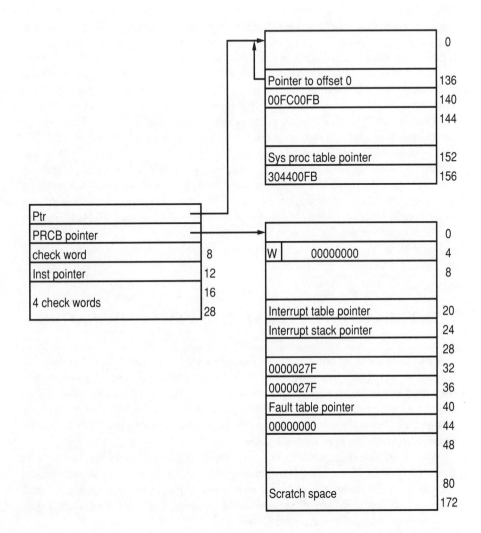

Figure 2-10: Initial memory image

Typically, the data structures in Figure 2-10 would be located in ROM (e.g., EPROM) in the system. However, notice that the PRCB contains one field that needs to be writeable: the 96-byte field labeled "scratch space." This field is used by the processor when a fault is generated, and during the processing of certain kinds of interrupts. Thus one has two options: (1) carefully locate the PRCB so that the first part falls in ROM and the scratch area falls in RAM, or (2) copy the PRCB into RAM right after initialization and use the REINITIALIZE-PROCESSOR IAC function to tell the processor of the new PRCB location.

Finally, it may useful to explain the presence of "funny" values in Figure 2-10, the existence of multiple data structures, and the existence of "blank" space in the data structures. There is one simple explanation: the higher architecture levels as described in the succeeding chapters. For instance, in the protected-architecture level (Chapter 4), more information is present in the PRCB, and the data structure above the PRCB becomes known as a segment table. Because of the objective of making each level of the architecture a pure subset of the higher levels (e.g., such that a program written for the core level could be executed on a protected-level processor with no change), the initialization data structures contain some information that would not otherwise be necessary.

INTER-AGENT COMMUNICATION MESSAGES

IAC messages perform functions similar to those of instructions, but they are defined in such a way that processors can send them amongst themselves on the bus (in a multiprocessing system). This allows, for instance, a program on processor A to send a message to processor B telling it to flush its instruction cache. Without this facility, processor A would need to generate an interrupt to processor B to tell a program in processor B to flush the cache.

Since IAC messages perform unique control functions that are not provided in instructions, their usage is important in single-processor systems. In this section the functions of IAC messages and their use in single-processor systems are defined. The next section discusses the usage of IAC messages that get transmitted externally over the bus and to other processors.

An IAC message consists of 1-4 words of information that are written to special memory-mapped addresses. The form of the 1-4 words of information is shown in Figure 2-11.

Msg type	Field 1	Field 2	0
Field 3			4
Field 4			8
Field 5			12

Figure 2-11: IAC message format

A processor can send itself an IAC message by writing the message to a special memory-mapped location, namely FF000010. The memory-mapping only occurs if one of the SYNCHRONOUS-MOVE instructions is used. A memory write to location FF000010 using one of these instructions does not cause a bus operation to occur; instead the data are interpreted by the processor as an IAC message, and the message causes some function to be performed by the processor. The function is performed synchronously (i.e., immediately after the SYNCHRONOUS-MOVE instruction). The IAC messages are shown below.

INTERRUPT
Field1 is an interrupt vector. Generates or posts an interrupt as described earlier.

TEST PENDING INTERRUPTS
Tests for a pending interrupt, as described earlier.

STORE SYSTEM BASE
Field3 is an address. Stores two words beginning at this address; the values are the addresses of the two initialization data structures.

PURGE INSTRUCTION CACHE
Invalidates all entries in the instruction cache.

SET BREAKPOINT REGISTER
(Described shortly.)

FREEZE
Puts the processor into the stopped state.

CONTINUE INITIALIZATION
Puts the processor at step 2 of the initialization sequence described earlier.

REINITIALIZE PROCESSOR
Field3, field4, and field5 contain, respectively, the addresses expected in memory locations 0, 4, and 12 during initialization. Puts the processor at step 4 of

the initialization sequence, using these addresses instead of those from memory.

EXTERNAL IAC MESSAGES

As discussed earlier, in addition to their use in performing certain implementation-dependent control functions, such as purging the instruction cache, IAC messages serve as a vehicle for processor control in multiple-processor systems. Much of this mechanism is carried out by the 82955 Bus Exchange Unit (or, in its absence, logic external to the processor providing the same interface); only those aspects of IAC messages that are pertinent to the processor are considered here.

Part of the physical address space is reserved for the transmittal of IAC messages, namely addresses of the form FFZZZZZ0, where ZZZZZ is the bit pattern aaaaaaaaaa00110ppppp. The a's represent the identifier of another processor, and ppppp represents the priority of the IAC message.

The transmittal of IAC messages via the BXU will be illustrated using Figure 2-12. Suppose processor A wishes to send an IAC message to processor B, telling processor B to freeze (stop), but only if processor B's priority is less than 30. A program in processor A would write a FREEZE IAC message to address FFZZZZZ0, putting processor B's 10-bit identifier and the priority value 30 in the address. This write operation passes onto the system bus, where BXU B recognizes processor B's identifier.* If BXU B's IAC buffer is empty, and if the priority in the address is greater than that in BXU B's priority register or if the priority in the address is 31, BXU B will store the IAC message data in its buffer and signal processor B that an IAC message has been received (via the IAC line). Otherwise, BXU B will reject the IAC message, which is communicated back to processor A via the processor's BADAC (bad access) pin.

Because write operations associated with external IAC messages can be rejected, external IAC messages should also be sent with one of the SYNCHRONOUS-MOVE instructions. These instructions do two things that normal STORE instructions do not do. First, they delay instruction execution until the operation is signaled as complete (unlike STORE instructions, which are pipelined). Second, they set the condition code indicating whether the operation was successful or not. This gives the sending program the ability to determine whether the IAC message was rejected.

---

*BXU's have a programmable register containing their attached processor's identifier, but the details of this are unimportant here.

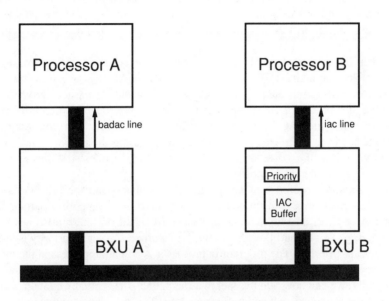

Figure 2-12: Sending IACs via the BXU

The above is the only interaction that programs have with external IAC messages. The remainder of the discussion below describes the handshaking that processors have with BXUs in order to receive and process the messages. This discussion is pertinent only to those desiring a better understanding of the handshaking, or to those wishing to design external logic to handle IAC messages instead of using the BXU.

Processor/BXU Interaction for Receipt of IAC Messages

To support the receipt of IAC messages, the BXU contains a control register (memory mapped at location FF000000), a four-word IAC-buffer register (memory mapped at location FF000010), and an IAC output pin. The control register contains the attached processor's current priority, and an indication of whether the buffer is empty or not.

The processor, by writing into the control register, can tell the BXU to (1) record a new priority, (2) record a new priority and mark the buffer as empty, or (3) mark the buffer as empty (without changing the priority). The processor is obligated to notify the BXU whenever the processor's priority changes. Thus, whenever an interrupt handler is called, or whenever a MODPC instruction is executed and the processor priority is changed as a result, the processor (the hardware, not software) writes the new priority into the BXU's control register. This occurs if the W (write-external-priority) flag in the PRCB is set. (In single-processor systems, one would normally give W the value 0.)

When the processor's IAC pin is asserted, the processor (the hardware) reads four words from address FF000010 (i.e., reads the IAC message from the buffer) and performs the indicated operation. Depending on the particular type of IAC message, the processor either immediately writes into the BXU's control register indicating that the IAC buffer is not empty, does so after the processing of the message has been completed (for IAC messages causing the processor to write into memory, such as the STORE-SYSTEM-BASE message), or does so along with a new priority (e.g., for the interrupt IAC message).

One important purpose of having the BXUs understand the priorities of their associated processors is in the transmittal of interrupts. Suppose a particular processor A (e.g., in an I/O subsystem) wishes to signal an interrupt to any of processors B, C, and D, but it needs to send it to a processor whose priority is less than the interrupt priority. The program in processor A can attempt to send the interrupt IAC message, in turn, to B, C, and D, stopping when someone accepts the message.

Finally, we can now add the fact that the 82955 BXU actually contains two IAC outputs and sets of control and buffer registers. This was done to allow one to place two processors behind a BXU.* One control/buffer register set is memory mapped at FF000000 and FF000010 as discussed before; the other set is mapped at FF000020 and FF000030. At reset time, the processor samples another pin that specifies whether it is processor "0" (i.e., should use BXU memory-mapped addresses FF000000 and FF000010) or processor "1" (i.e., should use the other addresses).

INTERRUPT CONTROLLER

The processor contains four interrupt pins and an interrupt control register that defines the meaning of the pins. The register is shown in Figure 2-13; it contains four bytes that define the meaning of the four interrupt pins (INT0-INT3). In its simplest form, the register defines the interrupt vector number that is used when an interrupt is signalled on the corresponding interrupt pin. The processor requires that the vectors are ordered by priority such that INT3's priority is greater than or equal to INT2's priority, and so on.

---

*To increase the number of processors in a system within the constraints of a maximum electrical load on the backplane bus.

| INT3 vector | INT2 vector | INT1 vector | INT0 vector |

Figure 2-13: Interrupt control register

The function of three of the four interrupt pins can be altered via special values in the interrupt control register. First, one has to decide whether to use pin INT0 as an interrupt pin or as the IAC-signaling pin; a zero value in bits 0-7 of the register makes it the latter. To extend the number of interrupt signals, pins INT2 and INT3 can be configured as handshake signals to an external interrupt controller (such as an Intel 8259A). This is specified by putting a zero value in bits 16-23 of the interrupt control register.

After reset, the value in the interrupt control register is FF000000 (meaning that the processor is configured for IAC message receipt and to function with an external interrupt controller). The interrupt control register is memory mapped at physical address FF000004, but can only be accessed in a memory-mapped fashion via the SYNMOV and SYNLD instructions.

SYNCHRONOUS LOAD/STORE INSTRUCTIONS

As discussed previously, the synchronous load/store instructions have several special purposes, such as reading/writing the interrupt control register, having the function of an IAC message performed by the processor, and sending an external IAC message. They can also be used to access any memory location in a synchronous fashion, determine (via the condition code) the success of the access, and perform noncacheable memory accesses.

For the SYNLD instruction, operand A is a register (or literal) containing an address and operand B is the destination register. If the address is FF000004, the value of the interrupt control register is stored in the destination and the condition code is set to 010. Otherwise, a one-word read is performed from the address, the processor waits for its completion, the value returned is placed in the destination register, and the condition code is set to 010 if the read was successful or 000 if unsuccessful (as denoted by the BADAC pin).

There are three corresponding store instructions, although they function as memory-to-memory moves. Operands A and B are registers (or literals) specifying the destination and source addresses respectively.*

For the SYNMOV instruction, the word at the source address is read. If the destination address is FF000004, the value is placed in the interrupt control register, and the condition code is set to 010. Otherwise, the value is written to the word specified by the destination address, the processor waits for completion of the write, and the condition code is set to 000 or 010 as in SYNLD.

The SYNMOVL instruction is similar, except that the read and write operations are for two words, and there is no test for FF000004.

The SYNMOVQ instruction is similar to SYNMOVL, except that four words are read and written. Also, if the destination address is FF000010, instead of performing the write operation the processor interprets the four-word value as an IAC message and performs the function indicated.

Any of SYNMOV, SYNMOVL, and SYNMOVQ can be used to send an external IAC message (the one used being dependent on the size of the message).

The processor contains an output pin named CACHEABLE. This pin can be used by the designer of an external cache to determine whether the current memory access is cacheable, or whether the memory locations accessed should not be brought into the cache. The 82955 BXU also uses this pin (as an input) for this purpose. In the context of the core architecture,** the pin is asserted for all memory accesses except for the read access of SYNLD and the write access of SYNMOV, SYNMOVL, and SYNMOVQ. This allows these instructions to be used for things like accessing memory-mapped I/O peripherals (where the behavior would be incorrect if the memory locations would be cached).

BREAKPOINT REGISTERS

The processor contains two special registers in which instruction addresses can be stored. The registers are used by Intel's in-circuit emulator, although they can also be used by software debuggers. When used by software debuggers, detection of a breakpoint (current IP is the same as the address in one of the registers), a trace fault occurs.

One can store into the registers by using the SET BREAKPOINT REGISTER IAC message, where field3 and field4 are the values stored into the breakpoint registers. Since instructions are aligned to word boundaries, bit 0 of the values is ignored and bit 1 indicates whether the breakpoint is enabled (0) or disabled (1).

---

*The instructions were defined as moves instead of stores because of a peculiarity in the implementation of the instruction decoder.

**The protected-architecture level provides for additional control of this pin.

REFERENCES

1. H. Kirrmann, "Data Format and Bus Compatibility in Multiprocessors", *IEEE Micro*, Vol. 3, No. 4, 1983, pp. 32-47.

2. C. A. Wiecek, "A Case Study of VAX-11 Instruction Set Usage for Compiler Execution", *Symp. on Architecture Support for Programming Languages and Operating Systems*, ACM, 1982, pp. 177-184.

3. D. R. Ditzel and H. R. McLellan, "Register Allocation for Free: The C Machine Stack Cache", *Symp. on Architecture Support for Programming Languages and Operating Systems*, ACM, 1982, pp. 48-53.

4. Digital Equipment Corp., *VAX-11/780 Hardware Handbook*, 1978.

CHAPTER 3

The Numerics
Architecture

The numerics architecture, the next level of the 80960 architecture, is an upward-compatible superset of the core architecture. The Intel 80960KB microprocessor implements the numerics architecture level.

The major difference between the numerics and core levels is the addition of floating-point arithmetic. The extensive set of floating-point arithmetic capabilities supports the popular IEEE floating-point standard.[1] Unlike many other implementations of the standard, which are highly incomplete and require significant support in software, the 80960 numerics architecture is an exceptionally complete implementation of the standard; the principal difference is the absence of operations to convert to/from decimal formats, which must be added via software routines. The 80960 architecture goes beyond the standard in many ways, for instance by containing many of the suggested functions in the appendix to the standard, by providing for mixed-length operations, and by providing a set of transcendental functions. In addition, the current 80960 processors excel in other ways that may not be immediately apparent to the user, such as the painstaking care that was taken to ensure that the transcendental functions have maximum accuracy.

This chapter is not a complete standalone discussion of the numerics architecture level. Instead, it builds on the previous chapter by discussing the differences between the core and numerics levels.

DATA TYPES

The numerics architecture adds four data types to those of the core architecture, as shown in Figure 3-1. The first three are 32-, 64-, and 80-bit floating-point representations, which are fully compatible with the corresponding data types in other Intel floating-point products, such as the 8087, 80287, and 80387.[2] The fourth is an ASCII decimal digit. The use of the terminology *real* to refer to floating-point values in the 80960 architecture stems from the FORTRAN and ALGOL languages.

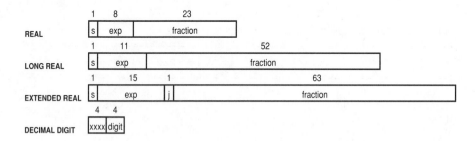

Figure 3-1: Additional data types

In the floating-point representations, *s* represents the sign of the value, *exp* represents the exponent (and its sign), and *fraction* the fraction or mantissa. The 80-bit representation contains another bit, *j*, that is logically part of the mantissa. All values are in base-2. The approximate range of numerical values is shown in Table 3-1.

Table 3-1: Approximate range of values

	Smallest	Largest
Real	-8.43×10^{-37}	-3.37×10^{38}
Long real	-4.19×10^{-307}	-1.67×10^{308}
Extended real	-3.4×10^{-4932}	-1.2×10^{4932}

It is important to note that many microprocessors and coprocessors implementing IEEE floating-point arithmetic do not include the extended-real (80-

bit) representation. However, because the extended-real data type has a range and precision far greater than the other types, it is invaluable for ensuring a high degree of accuracy in computations. For instance, in a language providing 32- and 64-bit data types, the 80-bit type can be used by the compiler for the intermediate results in evaluating expressions. Similarly, in mathematics libraries, the 80-bit type can be used for the intermediate calculations. Thus the presence of this data type significantly increases the probability that the final result of an expression or algorithm has maximum accuracy.

The encoding of the floating-point values is illustrated in Table 3-2.

Table 3-2: Floating-point value encodings

s	exp	j	fraction	Real	Long Real	Extended Real
x	max	1	nonzero	NaN	NaN	NaN
0	max	1	0	$+\infty$	$+\infty$	$+\infty$
1	max	1	0	$-\infty$	$-\infty$	$-\infty$
0	0	0	0	+0	+0	+0
1	0	0	0	-0	-0	-0
0	0<exp<max	1	frac	$+1.f \times 2^{exp-127}$	$+1.f \times 2^{exp-1023}$	$+1.f \times 2^{exp-16383}$
1	0<exp<max	1	frac	$-1.f \times 2^{exp-127}$	$-1.f \times 2^{exp-1023}$	$-1.f \times 2^{exp-16383}$
0	0	j	nonzero	$+0.f \times 2^{-126}$ (denormal)	$+0.f \times 2^{-1022}$ (denormal)	$+j.f \times 2^{-16382}$
0	0	1	frac			$+1.f \times 2^{-16382}$ (denormal)
1	0	j	nonzero	$-0.f \times 2^{-126}$ (denormal)	$-0.f \times 2^{-1022}$ (denormal)	$-j.f \times 2^{-16382}$ (denormal)
1	0	1	frac			$-1.f \times 2^{-16382}$ (denormal)
x	nonzero	0	x			reserved encoding

• Exponent value "max" is 255 (real), 2047 (long real), 32767 (extended real)
• The "j" column applies to only extended-real values

For those unfamiliar with the IEEE standard, striking aspects of the encodings are that there is an encoding representing a non-numerical value (NaN, meaning "not a number") and an encoding for infinity. The purpose of NaN and infinity is to allow computation to proceed under circumstances in which it would not in non-IEEE implementations, and at the same time giving the programmer the ability to detect certain types of computational errors. NaN and infinity are valid input values to instructions (e.g., to an add instruction), and NaN and infinity are possible results of instructions. $+\infty$ is treated as being greater than every finite number, and $-\infty$ is less than every finite number.

Also note that there are two zeros: +0 and -0. In almost all cases, they are treated as equivalent (same numerical value). However, the rounding mode used can influence whether a zero result is positive or negative, and certain operations that are influenced by the sign of their source values make subtle distinctions between +0 and -0.

For the encodings using the exponent and fraction to represent an arbitrary numerical value, the real and long-real formats assume the existence of an implicit 1 bit as the most significant part of the mantissa. This effectively gives one an extra bit of precision. In the extended-real format, this leading mantissa bit is explicit (the *j* bit). These values are known as *normals* or *normalized values*.

An important part of the IEEE standard is a concept of *gradual underflow*, where instead of a computation terminating abnormally when it produces a result that is smaller than the smallest representable normal number, the programmer has the option of letting it continue, albeit with a loss of precision. Such values are known as *denormals* (also known as subnormals), where the exp field is zero (i.e., largest negative exponent value), and the mantissa has one or more leading zeros (meaning that the processor needed to shift bits of precision off the end of the fraction). One should be able to see that progressively smaller denormals finally yield the value 0.

The last format, which only occurs in extended reals, is a reserved encoding. In the 8087, this is known as an *unnormal* and it has a numeric value. Unnormals are not supported in the 80960 architecture because they have little utility and add hardware complexity. However, the use of a value with the reserved encoding generates a unique type of fault, thus allowing one to emulate unnormal arithmetic in software if desired.

NaNs

The value known as NaN has several purposes. One purpose is the detection of the usage of uninitialized values. The programmer or compiler can give each variable that does not otherwise have an initial value the value NaN; then, as an option, one can cause any floating-point operation that has a NaN input to generate a fault. As another option, one can allow a computation to proceed despite the usage of NaN values, where an instruction that has a NaN input value produces a NaN as its result. Since the fraction field of a NaN is undefined, meaning that the program can assign unique NaN encodings to different variables, and since the fraction field of the NaN input is duplicated in the NaN result, one can look at the final NaN result of an algorithm and determine the source of the problem. Finally, as another option, instructions having no NaN inputs can produce NaN outputs under certain circumstances, without faulting. This allows an algorithm to be executed to completion, permitting the programmer to examine the results, some of which may be valid numeric values and some of which may be NaNs.

The architecture defines two types of NaN, as shown in Table 3-3. A *signaling* NaN (SNaN) is one where the most-significant fraction bit is 0; if an SNaN is used as an input to an arithmetic operation, an invalid-operation exception occurs. The exception can be masked, in which case no fault occurs and a QNaN result is produced, or the exception is unmasked, in which case a fault occurs. A *quiet* NaN (QNaN) is one where the most-significant fraction bit is 1; a QNaN input to an instruction always produces a QNaN result.

Table 3-3:   Types of NaNs

s	exp	j	fraction	Type of NaN
x	max	1	0 ... nonzero	SNaN (signaling)
x	max	1	1 ... nonzero	QNaN (quiet)
x	max	1	100 ... 00	QNaN (indefinite)

Within the QNaN class of NaNs, there is a specific encoding named *indefinite*. The indefinite QNaN is the QNaN produced in certain circumstances where an instruction has no NaN input but delivers a NaN result.

The floating-point instructions are divided into two classes: arithmetic instructions, where the instruction operates on its inputs as numerical values, and nonarithmetic instructions, where the instruction operates on its inputs as bit patterns. NaNs are only treated as such (as described above) for the arithmetic instructions.

As already stated, when an instruction has a NaN input, it produces an identical NaN output, except possibly for the highest fraction bit, which is set to produce a QNaN result. Actually, the rule is not quite this simple, because the architecture permits instructions with different-length operands, and because of the situation where the instruction has multiple NaN-valued input operands. The details for these situations can be found in the architecture reference manual.

ARITHMETIC CONTROLS

The numerics architecture adds a considerable number of control bits to the arithmetic-controls register, as shown in Figure 3-2. This figure also applies to the protected architecture level, since it does not add anything else to the arithmetic controls.

31   0

rc	nm	fp masks	---	fp flags	nif	---	iom	---	iof	--	astatus	ccode

BITS

rc	30-31	floating-point rounding control
nm	29	floating-point normalizing mode

fp masks
ixm	28	floating-point inexact mask
zdm	27	floating-point zero-divide mask
inm	26	floating-point invalid-op mask
ufm	25	floating-point underflow mask
ovm	24	floating-point overflow mask

fp flags
ixf	20	floating-point inexact flag
zdf	19	floating-point zero-divide flag
inf	18	floating-point invalid-op flag
uff	17	floating-point underflow flag
ovf	16	floating-point overflow flag

nif	15	no imprecise faults
iom	12	integer overflow mask
iof	8	integer overflow flag
astatus	3-6	floating-point arithmetic status
ccode	0-2	condition code

Figure 3-2: Arithmetic controls

Ten of the control bits consist of five flags and five mask bits. These are associated with five floating-point exceptions. (A sixth exception, called *reserved encoding*, has no associated control bits because it is meaningless to mask the exception.) If the mask is set, the occurrence of a floating-point exception causes the flag to be set and the instruction to generate a result. The flags are "sticky," meaning that the processor never implicitly clears them. If the mask is clear, the occurrence of a floating-point exception causes a floating-point fault to occur. For instance, if the program wants gradual underflow as discussed earlier, it would set the underflow mask; if the program wants to know immediately (via a fault handler) whenever an underflow occurs, it would execute with the underflow mask clear.

The arithmetic-status field is used by a few instructions as a place to store status information. The normalizing-mode flag, perhaps better named "don't trap on denormals," specifies whether denormals are to be treated as numbers (if set), or whether an instruction with a denormal input should fault. Normalizing mode exists in the 80960KB so that a complete implementation of unnormalized arithmetic can be provided in software, if desired. Since the major source of denormals is as the result of underflow (the other source being operands that are initialized to a denormal value), one would normally execute with the underflow mask and normalizing flag assigned the same value.

The rounding-control field specifies the type of rounding to be performed on the result of an instruction when the result is inexact. The setting are

00	- round to nearest
01	- round down (toward -∞)
10	- round up (toward +∞)
11	- truncate (toward 0)

Rounding toward the nearest value is the mode that is most-often used, since it produces a result with minimum error. In the case where the infinitely precise result is exactly halfway between two representable values, one might be inclined to think that the everyday "round up" rule would apply. However, since such a rule would introduce a bias into an algorithm, the architecture handles this by rounding to the nearest representable value whose least-significant fraction bit is 0.

Round up and round down are used in concert when one wants to perform interval arithmetic, or arithmetic with results that bound the error. For instance, to bound the result of A × B, one could first perform the operation with round-down in effect, and then repeat it with round-up in effect. This gives two results that are guaranteed to bound the infinitely precise result. By then carrying these values forth and using them both in further computations that use the result of A × B, one can produce a bounded final result.

The truncate rounding mode is the least useful. It can be used when performing floating-point to integer conversions, where the programming language specifies such conversions as truncating, but the 80960 architecture provides conversion instructions that always truncate rather than using the specified rounding mode.*

The 8087 has one mode, called *precision control*, that is absent from the 80960. Precision control allows one to specify that rather than producing an extended-real result with full precision, the rounding should occur in the extended-real result at the point in the fraction that has the same number of fraction bits as a real or long-real value. The 80960 architecture omitted this because it provides a facility far more useful: the ability for the operands of an instruction to be of different sizes. For instance, one can add two extended-real values and store the result in a real or long-real destination. When this is specified, the processor does not perform the addition by first adding the two extended-real values and producing an intermediate extended-real result (and possibly doing a rounding), and then converting the intermediate result to a real or long-real result, possibly with another rounding.

---

* These instructions are provided so that a FORTRAN compiler does not have to generate code that is constantly changing the rounding mode in the arithmetic controls.

This could produce an inaccurate result because of the possibility of two roundings. Instead, the processor performs such operations with only one rounding.

REGISTER MODEL

Floating-point operands may reside in the global and local registers, subject of course to the register-alignment rule (long reals must reside in an even/odd register pair, and extended reals occupy three registers, the first of which must be a multiple of four). In addition, the architecture defines four additional floating-point registers, as shown in Figure 3-3; these registers have a width of 80 bits. Floating-point instructions can refer to the local, global, or floating-point registers, but nonfloating-point instructions cannot refer to the floating-point registers.

80

	FP0
	FP1
	FP2
	FP3

Figure 3-3:   Floating-point registers

A secondary purpose of the floating-point registers is providing the mechanism for mixed-length arithmetic. This is accomplished by the following rule - if an operand of an instruction is a floating-point register, the operand is interpreted as an extended-real operand, regardless of the length specified by the instruction's opcode. For example, there are two floating-point add instructions: ADDR (32-bit add) and ADDRL (64-bit add). The following examples illustrate the use of the floating-point registers to provide mixed-length operations.

```
addr     r8,r9,r9       # 32-bit operands
addr     r8,r9,fp0      # 32-bit operands, result is 80 bits
addr     fp0,fp1,fp2    # 80-bit operands
addrl    r8,fp0,fp0     # add long real in r8 to extended real in fp0
```

There is no extended-real add instruction; thus arithmetic on extended reals is possible only when the extended reals reside in the floating-point registers. Also, although the architecture permits the mixing of 32-bit/80-bit and 64-bit/80-bit operands, the instruction set does not permit mixing of 32-bit/64-bit operands. To add a real to a long real, one needs to move one of the operands into a floating-point register (which converts it to an extended real) and then use the ADDR or ADDRL instruction.

Why have the floating-point registers? One reason is the previous discussion: they provide the means for expressing mixed-length operands. The major reason, however, is performance. First, the array of local and global registers needs to be located within the integer execution unit of the chip, but the floating-point registers are located within the floating-point execution unit. Thus, the floating-point registers are "closer" to the floating-point ALU, meaning less time is needed to access them. Second, at least in the current implementation, the data path in the integer execution unit is 32 bits wide. Therefore, fetching a long-real operand from the global or local registers costs two clock cycles, and fetching an extended-real operand from there would cost three clock cycles.

Why four floating-point registers? This was one of the many tradeoffs between architecture and chip size; we could not afford the space for more. Compilers use the floating-point registers primarily for intermediate results during the evaluation of an expression, and four registers are reasonable for this. Of course, one could write algorithms to make more direct use of the floating-point registers, but studies of the inner loops of many algorithms showed that four registers are usually adequate. The encoding of the floating-point instructions is such that additional floating-point registers could be added in future implementations.

INSTRUCTION FORMAT

The floating-point instructions have the same format as the instructions in Chapter 2. All of the floating-point instructions fall into the REG-format class (Figure 2-2); the only difference is how the floating-point registers and literals are encoded.

In the integer instructions, the m bit associated with each of the three operands specifies whether the operand is a register or a literal value (0-31). In the floating-point instructions, the m bits are interpreted differently. If 0, the interpretation is the same as for the integer instructions (i.e., the register field names one of the 16 local or 16 global registers. If 1, register-field (source1, source2, or src/dest) values of 00000-00011 denote registers fp0-fp3, and the values 10000 and 10110 denote literal values +0.0 and +1.0 respectively. Of course, all of this detail is hidden from the assembly-language programmer by the assembler.

As mentioned earlier, sufficient encoding space remains to add more registers or literals in future implementations. An analysis of floating-point programs showed that the constants +0.0 and +1.0 appear frequently (with all other constant values being in the noise), which is why they were included as floating-point literals. Not having them available as literals would require that they be loaded from memory for each use (or, be allocated in registers, reducing the number of available registers).

INSTRUCTION SET

The floating-point instructions in the numerics architecture are listed in Table 3-4. As discussed before, most of the operations have a real and long-real instruction, and extended-real operands can be used by both via a floating-point register.

Each instruction is considered either *arithmetic* or *nonarithmetic*. Arithmetic instructions are sensitive to the value of the floating-point operand(s) and can generate a floating-point fault. Nonarithmetic instructions treat their operands as bit patterns and do not generate floating-point fault.

Unlike most of the other operations, there is an extended-real form of the move instruction (MOVRE). The rationale is that an instruction is needed that can refer to an extended-real operand in the local/global registers (in order to move an extended-real operand into and out of the floating-point registers). Only the MOVRE move instruction is nonarithmetic; the rationale is that without this, there would be no way to put an SNaN into a floating-point register.

The REMAINDER instructions perform modulo division of one input by the other, producing a true remainder. The remainder is always exact providing that the destination is not shorter than the inputs. The remainder is $B - N \times A$, where N is the integer nearest to the exact value B/A and where $abs(N) \leq abs(B/A)$. Calculation of the remainder can take an extraordinary amount of time because it is done by repeated subtraction. This does not affect interrupt response time, because the architecture provides a mechanism for suspending and later resuming interrupted instructions. The 8087 does not have such a mechanism, and thus only provides a partial-remainder instruction, which must be enclosed in a software loop to find the true remainder.

The REMAINDER instruction also stores some status bits in the arithmetic-status field of the arithmetic controls. This status is needed to adjust the result of the instruction if one wants a remainder conforming exactly to the IEEE standard.* The details of the status bits can be found in the architecture reference manual.

---

* The reason for the difference is that REMR and REMRL produce a result between 0 and A, which is usually the remainder one wants instead of the IEEE-defined remainder, which is between -A/2 and A/2.

Table 3-4: Floating-point instructions

Mnemonic(s)		Function	Result	Arithmetic	Sets Ccode
movr	movrl	move	A	X	
movre		move	A		
addr	addrl	add	B + A	X	
subr	subrl	subtract	B - A	X	
mulr	mulrl	multiply	B × A	X	
divr	divrl	divide	B ÷ A	X	
remr	remrl	remainder	see text	X	
scaler	scalerl	scale	$B \times 2^A$	X	
roundr	roundrl	round to integral value	int(A)	X	
sqrtr	sqrtrl	square root	$\sqrt{A}$	X	
sinr	sinrl	sine	sin(A)	X	
cosr	cosrl	cosine	cos(A)	X	
tanr	tanrl	tangent	tan(A)	X	
atanr	atanrl	arctangent	atan(B/A)	X	
expr	exprl	exponential	$2^A - 1$	X	
logbnr	logbnrl	get exponent	see text	X	
logr	logrl	logarithm	$B \times \log_2(A)$	X	
logepr	logeprl	logarithm	$B \times \log_2(1+A)$	X	
classr	classrl	classify	see text		
cpysre		copy sign	see text		
cpyrsre		copy reversed sign	see text		
cmpr	cmprl	compare	A ? B	X	X
cmpor	cmporl	compare ordered	A ? B	X	X
cvtri	cvtril	convert to integer	int(A)	X	
cvtzri	cvtzril	convert truncated to int	int(A)	X	
cvtir	cvtilr	convert to real	real(A)	X	

The following code uses REMRL in a function that implements a 64-bit remainder function matching the standard.*

```
# z = ieee_rem(x, y)
#
# z is in g0,g1
# x is in g2,g3
# y is in g4,g5
#
# assumes g2..g5 do not need to be preserved
#
ieee_rem:
        remrl     g4,g2,g0
        modac     0,0,r4       # Get AC in r4
        bbc       4,r4,rem2    # QR=0 means g0 < y/2 means z = g0
        bbs       3,r4,rem1    # QR=1,QS=1 means g0 > y/2 means z = g0 - y
        bbc       5,r4,rem2    # QR=1,QS=0,Q0=0 means g0 = y/2 means z = g0
                               # QR=1,QS=0,Q0=1 means g0 = y/2 with odd
                               # quotient means z = g0 - y
rem1:   chkbit    31,g5
        alterbit  31,g5,g5     # give y same sign as result
        subrl     g4,g0,g0     # g0 (z) = g0 - y
rem2:   ret
```

The SCALE instructions provide a fast way to multiply or divide a value by a power of 2. The ROUND instructions round the input value to an integral value (using the rounding mode in effect) and return this value as a floating-point value. The SQRT instructions compute the square root of the operand. In adherence to the IEEE standard, $\sqrt{-0} = -0$.

The next set of instructions in the table perform a set of frequently used transcendental functions. A great deal of care was taken with the implementation of these functions in terms of speed, accuracy, and adherence to certain mathematical properties. For instance, an internal constant π is needed in the algorithms for some of these instructions, and this constant is represented with 66 bits of precision (two more than in the extended-real format). In general, these functions produce an accurate result (except for rounding) when the destination is real or long real, but may be inaccurate in the last few bit positions when the result is extended real. The functions are carefully implemented to be monotonic (e.g., if $x > y$, $\sin(x) \geq \sin(y)$).

Note that the ATAN instructions calculate the arctangent function of the *ratio* of the two input values. This was done for ease of using the ATAN instructions with

---

*Courtesy of James Valerio.

standard identities to calculate other standard functions, such as arcsine and arc-cosine.

The EXPONENTIAL instructions calculate 2^x-1. The reason for this instead of 2^x is that the former allows the result to be produced with more accuracy when x is close to zero; if one wants 2^x, one simply adds 1 to the result. One is usually interested in calculating exponentials for bases other than 2. This is easily accomplished with the EXPR and EXPRL instructions. For instance, $10^x = 1+(2^{xc} - 1)$, where c is $\log_2(10)$. Also, $y^x = 1+(2^{xz} - 1)$ where z is $\log_2(y)$.

To reduce the complexity of the on-chip exponential algorithm, there is a range restriction on x: $-0.5 \leq x \leq 0.5$. One can calculate 2^x for any x by an algorithm that does roughly the following. The rounding mode is set to round to nearest, and the ROUNDR/ROUNDRL instruction is used on x (producing an integral value m). When x is subtracted from the integral result of the round instruction, we have two values, one is the integral value m, and the other is a value n in the above range (one half or less), where $m+n = x$. This means that $2^x = 2^{m+n} = 2^n \times 2^m$. 2^n can be obtained from the EXP instruction, and 2^m, since m is an integral value, can be obtained by converting m to an integer and using the SCALE instruction.

The LOGBNR and LOGBNRL instructions return a floating-point value equal to the unbiased exponent (i.e., exp minus 127, 1023, or 16383) of the input value. As is the case with all instructions, the behavior is well-specified for all possible input values. For instance, logbn(NaN) = NaN, logbn(+∞) = logbn(-∞) = +∞, logbn(0) = -∞ and causes a divide-by-zero exception, and logbn of a denormal returns the exponent that the value would have if its format had an unlimited exponent range.

The LOGR and LOGRL instructions compute B $\times \log_2$(A). The reason for the extra multiplication is that most uses of these instructions entail a multiplication. Of course, if the instructions did not include the multiplication, one could use a MULTIPLY instruction after the LOGR/LOGRL instruction. However, this would entail a rounding of $\log_2$(A). The LOGR/LOGRL instructions perform the multiplication on the unrounded calculation of $\log_2$(A), and thus provide more accuracy.

Logarithms in other bases are easily computed by assigning B a constant value. For instance, $\log_{10}(x) = \log_{10}(2) \times \log_2(x)$.

The LOGEPR and LOGEPRL instructions compute B $\times \log_2$(1+A). They produce a result that is more accurate than the above instructions when A is small (close to zero). An example would be in the computation of daily compounded interest, where A is the daily interest rate. Like the EXP instructions, LOGEP has a range restriction: $-1/\sqrt{2} \leq 1+A \leq \sqrt{2}$.

The CLASS instructions store a value in the 4-bit arithmetic-status field in the arithmetic controls indicating the type of value of the operand. This operation is one of the recommended functions in the appendix of the IEEE standard, and is useful in implementing software functions that adhere to the standard. The value stored in the arithmetic-status field is shown in Table 3-5.

Table 3-5: Result of the CLASS instructions

Status	Operands
s000	Zero
s001	Denormalized number
s010	Normal nonzero finite number
s011	Infinity
s100	Quiet NaN
s101	Signaling NaN
s110	Reserved encoding

s is the sign bit of the operand

As an example, the following sequence branches if the 32-bit operand in register r4 is not a finite number.

```
classr      r4
modac       0,0,r5        # Put AC in r5, arith status is bits 3-6
shro        3,r5,r5       # Arith status now bits 0-3
and         7,r5,r5       # r5 = bits 0-2 of arith status
cmpobge     r5,3,errcase  # branch if status was not s000,s001,or s002
```

The IEEE standard permits the implementation to decide whether operations on the sign of an operand, namely abs(x) and -x, are arithmetic or not. We chose the latter (not). Thus, the fastest way to perform these operations is via the bit instructions. To form abs(x) on a floating-point value in a local or global register, one uses the CLRBIT instruction (on the sign bit). To form -x, one uses the NOTBIT instruction. However, the bit instructions, not being floating-point instructions, cannot refer to the floating-point registers. To solve this problem, the CPYSRE and CPYRSRE instructions are provided. They assume an extended-real operand and, although they can be used on operands in the local/global registers, their intention is for use on operands in the floating-point registers. CPYSRE puts the A operand in the destination, giving it the sign of the B operand. CPYRSRE puts the A operand in the destination, giving it the inverse of the sign of the B operand. Their typical uses are illustrated below.

```
cpysre      fp1,1.0,fp1    # fp1 = abs(fp1)
cpyrsre     fp1,fp1,fp1    # fp1 = -fp1
```

The next set of instructions are comparisons. The important thing to realize about floating-point comparisons is that there are four mutually exclusive relationships instead of three: less than, equal, greater than, and *unordered*. Unordered is the relationship that holds if any of the two operands is a NaN. Because of this, one cannot invert comparison relationships in the customary sense. For instance, if it is false that two operands are equal, it is not necessarily true that they are unequal or that one is greater than the other; they can be unequal or unordered.

Two types of comparison instructions are provided: one of which results in an exception if the unordered relationship is present, and the other of which sets the condition code to a unique value for unordered. The CMPR/CMPRL instructions set the condition code to 100, 010, or 001 if the relationship is less, equal, or greater, respectively, and generates an exception for the unordered case. The CMPOR and CMPORL instructions set the condition code to 100, 010, 001, or 000, the last being the case for unordered. One should now be able to see the purpose of the BRANCH, TEST, and FAULT instructions (in Chapter 2) that deal with ordered and unordered. For instance, the BNO instruction branches if the condition code is 000. The BO instruction branches if any condition code bit is 1 (i.e., if the operands are ordered).

As is the case for most of the instructions, the infinite values are valid operands. For instance, +∞ is greater than any other number (other than +∞).

The last set of floating-point instructions are conversion operations between floating-point and integer values. CVTRI converts a floating-point value to a 32-bit integer, rounding the input value to an integral value using the current rounding mode. CVTRIL produces a 64-bit integer. CVTZRI and CVTZRIL do the same, except they truncate (round toward zero) instead of using the current rounding mode. These exist because some programming languages define the conversion as truncating, and not having these instructions would require the compiler to change repeatedly the rounding mode in the arithmetic controls. CVTIR and CVTILR convert a 32-bit integer or 64-bit integer to a floating-point value.

No conversions exist for long (64-bit) reals, since one can do this easily by using the floating-point registers. For instance, to convert a long real to 32-bit integer, one would use a MOVRL instruction to move the value into a floating-point register (making it an extended real), and then use the CVTRI instruction, naming the floating-point register as the source operand.

There are no load and store instructions for the 80-bit floating-point registers, meaning that one must load and store via the 32-bit local and global registers. There were several reasons for this. First, the floating-point registers do not "fit" well within the load and store instructions, both from an encoding and semantic point of view; solving this would have required additional instructions. Second, supporting direct loads into the floating-point registers would have complicated the register-scoreboard logic. Third, the first-generation chip design had no direct path between the external bus controller and the floating-point registers, meaning that the existence of floating-point-register loads and stores would have added additional control sequencing logic to the chips.

Finally, seven non-floating-point instructions are added in the numerics architecture level. They are listed in Table 3-6.

Table 3-6: Non-floating-point numerics-architecture instructions

Mnemonic	Function	Sets CCode
dmovt	decimal move and test	X
daddc	decimal add with carry	X
dsubc	decimal subtract with carry	X
concmpo	conditional compare (unsigned)	X
concmpi	conditional compare	X
emul	extended multiply	
ediv	extended divide	

The first three instructions provide a basic set of operations for use with ASCII-encoded unpacked decimal data. The DMOVT instruction is similar to the MOV instruction (i.e., moves a 32-bit value), but it also tests the value for the pattern XXXXXX3M, where M is 0000-1001. In other words, it tests for low-order byte to determine if it is a valid ASCII digit. If it is, the condition code is set to 000; otherwise the condition code is set to 010.

The DADDC instruction is intended for iterative addition of two decimal strings. The middle condition-code bit is both an input and output. It adds two values of the form XXXXXXXM plus the condition-code bit, producing a result of the form XXXXXXXM and setting the condition code to 010 if a decimal carry-out occurred, or to 000 otherwise. For instance, if the two source registers have the values 30303037 (A) and 31313138 (B) and the condition code is initially 000, it assigns the value 31313135 to the destination register and sets the condition code to 010.

One would add two four-digit decimal numbers by a loop containing a DADDC instruction and a ROTATE instruction.* One would add larger strings by using additional registers or loading and adding four bytes repeatedly.

The DSUBC instruction is similar, except that it performs a subtraction and uses the middle condition-code bit as a borrow flag.

The conditional compare instructions are provided for two-sided range comparisons, such as testing whether X is between Y and Z. The condition code is both an input and output. If the condition code has the value 1xx, the instruction

---

*However, one needs to be careful to preserve the condition code from one iteration to the next.

acts as a no-op. Otherwise, it sets the condition code to 010 if A≤B, or to 001 if A>B. If one wants to branch if the value in register r4 is not from 5 to 15, one could write

```
cmpo       r4,5
concmpo    r4,15
bne        out_of_range
```

Without the CONCMP instructions, one would need to use an additional branch instruction. In pipelined machines, such as the 80960KB, branch operations cause pipeline breaks, and thus extraneous branches should be avoided.

The EMUL and EDIV instructions provide support for software-created operations on signed and unsigned integers of widths greater than 32 bits. The EMUL (extended multiply) instruction produces a 64-bit result, and the EDIV instruction divides a 64-bit dividend by a 32-bit divisor, producing a 64-bit quotient.

INSTRUCTION RESUMPTION

The presence of instructions with long execution times could present a problem with respect to interrupt responsiveness. The architecture specification deals with this by giving the implementation three options for handling instruction execution in the presence of an interrupt:

1. Instruction completion. No action is taken on the interrupt until the current instruction(s) completes. This is the obvious choice for most instructions.

2. Instruction restart. The instruction is terminated and the instruction pointer is not advanced.

3. Instruction suspension. The instruction is suspended and its intermediate state is saved, such that execution can be resumed later from the point of interruption.

The 80960KB uses all three techniques. For certain long-running instructions (e.g., square root), the instruction is simply aborted if an interrupt occurs. However, four instructions (REM, SIN, COS, and TAN) can have, in certain circumstances, extremely long execution times such that if they were handled by instruction restart, their execution might never be completed in an environment with continuous frequent interrupts. For the REM instruction, the reason was explained earlier; the production of the exact remainder involves repeated subtractions. For SIN, COS, and

TAN, the execution time is reasonably short when the input value is small (e.g., $\pi/2$ or less), which is the typical case. However, if the input value is large, the algorithms must first reduce the value by invoking the remainder operation internally.

The technique of instruction suspension entails an addition to the process controls and to the mechanisms of calling an interrupt handler and returning from an interrupt. These are discussed below.

One flag is added to the process controls, as shown in Figure 3-4. The resume flag is used to remember if a suspended instruction needs to be resumed.

```
31                                                                              0
 _____
|              |          |  |  |  |  |  |   |   |  |       |   |   |
| internal state |  priority  |--| 0|st| 0|--|tfp|res| 0|  ---  | em| t |
|_____|_____|__|__|__|__|__|___|___|__|_____|___|___|
```

	BITS	
internal state	21-31	internal state - implementation dependent
priority	16-20	priority (0-31)
0	14	must be zero
st	13	state (0 = executing, 1 = interrupted)
0	12	must be zero
tfp	10	trace fault pending
res	9	resume suspended instruction
0	8	must be zero
em	1	execution mode (0 = user, 1 = supervisor)
t	0	trace enable

Figure 3-4: Process controls

The operations that occur when an interrupt handler is to be called become more complex, in that when an interrupt causes an instruction to be suspended, a *resumption record* is stored on the stack. The resumption record contains information used by the processor to resume a suspended instruction. Since the information saved to resume instructions is expected to vary from implementation to implementation, the nature of the resumption information is not specified by the architecture. Therefore, instead of the three steps defined in Chapter 2, the following occurs when an interrupt handler is called.

1. The value of the process controls is saved. If the interrupt occurred within an instruction that was suspended as a result, the *resume* flag is set in the saved process controls.

2. In the process controls, the state is set to interrupted, the execution mode is set to supervisor, the trace flags are cleared, and the priority is set to that of the interrupt.

3. A call operation is done to the procedure specified in the interrupt table. If the processor was in the interrupted state prior to the interrupt, the value of SP used in the call operation is SP+16+M; otherwise the value of SP used is ISP+16+M, where ISP is the interrupt stack pointer, and M, a value in the range 0-48, is the size of the resumption record needed. The return status in the new r0 register is set to 111.

    Finally, prior to starting execution of the interrupt handler, the saved process-controls word is stored on the stack at address FP-16, and the arithmetic-controls word is stored at address FP-12, and if the *resume* flag is set in the saved process controls, a resumption record is copied to the stack at address FP-16-M.

At the end of the interrupt return operation, the *resume* flag in the restored process controls is tested. If set, the processor loads the resumption information from the stack, clears the flag, and uses the information to resume execution of the next instruction (the interrupted instruction).

EXCEPTIONS

As mentioned earlier, some conditions encountered during a floating-point instruction result in an exception. An exception is not necessarily an error; it represents an exceptional condition which the program can choose to ignore (meaning that the instruction produces some default result) or to handle via a call to a fault handler. An exception can be ignored by setting its mask bit in the arithmetic controls. When a masked exception occurs, the processor sets the corresponding sticky bit in the arithmetic controls, meaning, for instance, that a program can choose to mask underflow exceptions and then test the underflow flag at the end of the program to see if any underflows occurred.

Because of the above, discussion of the exceptions requires defining the result returned by the instruction if the exception is masked.

Invalid Operation Exception

This exception occurs in a large number of cases where the value of an operand prohibits the returning of a meaningful arithmetic result. Some of the cases are

- Operand is an SNaN
- $+\infty + -\infty$
- $0 \times \infty$
- $0/0$ or ∞/∞
- sqrt(x) where x < -0
- cvtri(NaN)
- $\sin(\infty)$
- Operand of the CMPOR instruction is a NaN

If the exception is masked and the instruction produces a floating-point result, the result is the indefinite QNaN; if the instruction produces an integer (i.e., is a conversion instruction), the maximum negative integer value is produced.

Division by Zero Exception

This exception occurs for x/0, where x is finite and nonzero. It also occurs in the LOGBN instruction where the input is zero. If the exception is masked, the result returned is a correctly signed ∞.

Overflow Exception

This exception occurs whenever the result of an instruction, assuming its exponent range is unbounded, exceeds the magnitude of the largest finite number of the destination data type. If the exception is masked, the result returned depends on the current rounding mode. Round to nearest produces ∞ with the sign of the unbounded result. Round down produces the largest positive finite number for positive overflow and $-\infty$ for negative overflow. Round up produces the most-negative finite number for negative overflow and $+\infty$ for positive overflow. Round to 0 (truncate) produces the largest finite number with the sign of the unbounded result.

When the exception is not masked, no result is stored in the destination, and an extended real value (the unbounded result) is stored in the fault record. If the unbounded result has an exponent that exceeds the range of the extended-real format, unbounded result is divided by 224576, this value is stored in the fault record, and a flag is set in the fault record indicating that the value has been "bias adjusted." If the overflow occurred in the scale instruction and the unbounded result is even too large for its exponent to be bias-adjusted as described above, a properly signed ∞ is stored in the fault record.

Underflow Exception

The underflow exception is unusual in that the condition under which the exception occurs depends on the setting of the underflow mask. The underflow exception occurs when the unbounded result is nonzero and would lie between -2^{emin} and 2^{emin} after rounding. The value emin is -126 for real destinations, -1022 for long real, and -16363 for extended real. The underflow exception also occurs when the mask is set and the result is different from what it would be if both the exponent range and precision were unbounded.

When the exception is masked, the possible results returned are 0, -2^{emin}, 2^{emin}, or a denormal. When the exception is not masked, the effect is similar to overflow. If the exponent in the fault record needs to be bias-adjusted, the unbounded result is multiplied by 224576. If this is insufficient (in the case of massive underflow in the SCALE instruction), a properly signed zero is stored in the fault record.

Inexact Exception

The inexact exception occurs if the result is inexact (is unequal to the infinitely precise true result), or if the overflow exception occurred but was masked. When the inexact exception is masked, the result returned is the rounded result or the result as defined above for the overflow exception. When the exception is not masked, and if no other exception occurred, the rounded result is stored in the instruction's destination. When the exception is not masked and overflow or under-flow also occurred, no result is stored in the destination, and an extended-real value is stored in the fault record as described for the overflow and underflow exceptions.

Reserved Encoding Exception

This exception occurs when an extended-real operand has the reserved encod-ing, or when an operand is a denormal and normalizing mode is not in effect. This exception has no corresponding mask (i.e., a fault always occurs).

FAULTS

The numerics architecture extends the fault architecture of the core architecture by adding another fault type and expanding the size of the fault record.

The added fault type is the floating-point fault. Its type encoding is 4. Its sub-type encoding (see Table 2-8) is

xxxxxxx1	Overflow
xxxxxx1x	Underflow
xxxxx1xx	Invalid operation
xxxx1xxx	Division by zero
xxx1xxxx	Inexact
xx1xxxxx	Reserved encoding

The subtypes are encoded as bits because there is one situation where two can occur simultaneously.

In the fault table (Figure 2-7), the entry for the floating-point fault is at offset 32.

The floating-point fault is an imprecise fault. If the fault handler wants to resume execution from the point of the fault, one should set the *nif* flag in the arithmetic controls.

Because of the need to store additional information in the fault record, the fault record is expanded, as shown in Figure 3-5. The end of the fault record (address of faulting instruction) always appears at the same place, namely, at address FP-4 from the point of view of the fault handler.

Fault Data				0
reserved				12
Process Controls				16
Arithmetic Controls				20
FFlags	FType		FSubtype	24
Address of Faulting Instruction				28

Figure 3-5: Fault record

As explained earlier, in the case of underflow and overflow floating-point faults, additional information is stored in the fault record. In the fault-data field, an adjusted extended-real result is stored. This result is rounded to the precision of the destination of the instruction that faulted. The *FFlags* field contains two bits that allow the fault handler to reconstruct the true result from the adjusted result. Bit 1 of this field is set if the adjusted result has been bias-adjusted (because the true exponent is outside of the exponent range of an extended real). Bit 0 of this field is meaningful only if the inexact exception occurred with the underflow or overflow. If set, it indicates that the result was rounded up; if clear, the result was rounded down.

What do Floating-Point Fault Handlers do?

The most-obvious thing that one might do in a fault handler is to describe the nature of the error in an error message and terminate the program, possibly interacting with the compiler's run-time environment when doing so. However, there are other things that one might do in the fault handler.

One possibility is to have the fault handler record the exception (e.g., type, instruction address, instruction operands), construct the default result (the result that would have occurred had the exception been masked) and put it in the faulting instruction's destination, and then return to the following instruction. At the end of the program, the user could examine the recorded exceptions to decide whether they invalidate the computation or not.

Another possibility in an interactive environment, such as a workstation, is for the fault handler to communicate with the user, such that the user can choose to stop the program or have the default result or user-specified result returned and continue execution.

Intel provides a software floating-point support library that provides additional high-accuracy floating-point functions as well as handling exceptions. If the user chooses not to mask a particular exception, and that exception occurs, the fault handler can be requested to call a user-supplied routine, passing to the user routine the instruction's input operands and what the default result would have been had the fault been masked. The user routine can then make the decision of what to do (e.g., terminate, record and continue, return the default or another result and continue).

Finally, in small, well-understood programs, there is another possibility in the event of overflow and underflow, one that can extend the exponent range of a computation even beyond that in extended reals. When an overflow or underflow occurs and gives a bias-adjusted result to the fault handler, one can record the exponent adjustment in a counter associated with the result operand, assign the bias-adjusted value to the instruction's destination, and continue. At the end of the program, one can use the values in the counters to adjust the final results of the program.

REFERENCES

1.   IEEE, *ANSI/IEEE Std 754-1985: IEEE Standard for Binary Floating-Point Arithmetic*, 1985.

2.   J. F. Palmer and S. P. Morse, *The 8087 Primer*, Wiley, New York, 1984.

CHAPTER 4

The Protected Architecture

The protected architecture, the next level of the 80960 architecture, is an upward-compatible superset of the numerics and core architectures. The Intel 80960MC microprocessor implements the protected architecture level.

The protected architecture contains four sets of features that are not present in the numerics level: address translation to provide virtual memory and memory protection, process (task) management, multiple-processor support, and additional instructions. Although one normally associates address translation with systems that need virtual memory and paging (e.g., workstations, computer systems), it has wider applicability; address translation provides a mechanism to isolate the memory spaces of processes and to protect the operating system from application programs. In fact, some of the earliest users of the product line, who used the processor in embedded avionics systems, selected the 80960MC instead of the 80960KA and 80960KB because of its memory protection capabilities. Also, many of the additions in the 80960MC are particularly useful in supporting the run-time environment of the Ada language.

This chapter is not a standalone discussion of the protected architecture level. Instead, it builds on the previous chapters by discussing the additions in the protected architecture.

PROCESSES AND ADDRESS SPACES

One major addition in the protected architecture is the concept of a process (or task). The state of a process is described in a data structure defined by the architecture called a *process control block*, or PCB. One of the key attributes of a process is its specification of a 2^{32}-byte virtual address space. Considerable flexibility is present in the association of a process to a virtual address space, allowing, for instance, each process to have a private address space, multiple processes to share a single address space, and multiple processes to share parts of their address space.

The relationship between processes and address spaces is shown in Figure 4-1. Each process has an address space that is divided into four *regions*. Each process, by means of mapping tables associated with its PCB, specifies regions 0-2. The architecture defines that region 3 is shared among all processes because, independent of the process that is running, the system needs a uniformly addressable place for things that are not associated with a process, such as the interrupt stack and interrupt handlers. Depending on the type of system being designed, region 3 might be used for other things as well, such as containing the operating system and holding data or code shared among all processes. Each region is defined by a separate set of mapping tables (thus, two processes have the same address space if their PCBs point to the same mapping tables, and two processes can share a single region by sharing its mapping table).

The region boundaries are transparent to a program in an address space, and the address space looks like a linear address space referenced with 32-bit addresses. However, an operand cannot span region boundaries.* In almost all cases, this is not a problem because operating systems and loaders typically allocate memory by mapping a process's stack into one region, its code into another, and its data into the third. However, if one wishes to have operands span the region boundaries, one can do so by a clever ploy that involves using a set of overlapping mapping tables for the regions.

Why divide the address space into regions? The regions provide a way to share all or parts of the address space among processes (although the same can be accomplished by sharing all or parts of the underlying mapping tables). Another reason for dividing the address space is to provide for a process-independent part of the address space (region 3), as described above. If region 3 were not defined as system-wide, one could achieve the same by software convention, by having all processes point to the same mapping tables for region 3. However, performance benefits result from having one part of the address space remain unchanged across process switches.[1] For instance, the processor contains a TLB (translation lookaside buffer) in which mapping information is maintained for most recently referenced pages (a page is a 4k-byte chunk of the address space). Addresses that "hit" the TLB are translated in a single cycle. Normally, when the processor switches from one process to another, it must purge the TLB. However, it does not purge any entries associated with region 3, since the architecture defines region 3 as common to all processes.

---

*There is no inherent reason for this, but the 80960MC makes this assumption to reduce complexity.

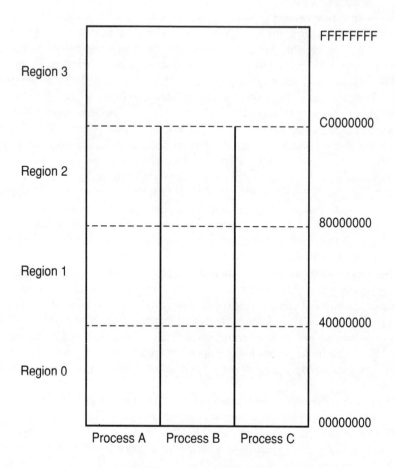

Figure 4-1:   Process address spaces

The discussion above makes the assumption that the processor is being used with address translation enabled. When address translation is disabled, a process still sees a 2^{32}-byte address space, but the address space is the physical address space, and thus all processes reside by definition in the same address space.

REGION MAPPING

Virtual addresses are translated to physical addresses by converting the upper 20 bits and leaving the lower 12 bits intact. In other words, the upper 20 bits of the virtual address are translated to the physical address of a page, where a page is a 4k-byte unit of the address space on a 4k-byte boundary. Since each region has its own mapping tables, the upper 2 bits of the virtual address denote a region, the next 18 bits denote a page within that region, and the 12 low-order bits denote a byte within that page.

For flexibility, a region can be mapped by 0, 1, or 2 levels of tables, as shown in Figure 4-2. In the first case, the region is defined as a single page; in this case STE, a descriptor for the region, points directly at a physical page. In the second case, the region is defined by a page table, which points at a maximum of 1024 physical pages. In this case, the region can be mapped into 4M bytes of physical memory. In the third case, the region is defined by a page table directory, whose entries point to individual page tables, which in turn point to individual pages. The third case is the mapping that must be used for regions of maximal size.

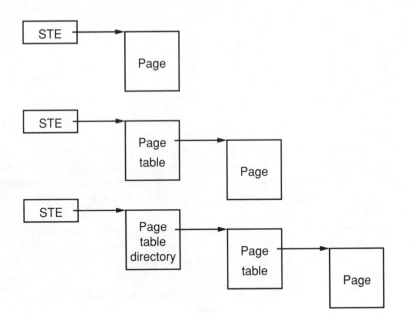

Figure 4-2:   Mapping regions to pages

Why have large, 4k-byte pages? In contrast, the page size of the VAX architecture is only 512 bytes. The primary reason is trends in memory and I/O technology. Assuming one is using paging (the swapping by software of pages between main memory and an I/O device on demand)*, the optimal page size depends on the memory size, the access time of the device (e.g., disk) used to hold the swapped-out pages, and the bandwidth (data rate) of transfers between this device and memory. Everything else being constant, the trend to larger memory sizes points to larger page sizes. The trend to longer access or seek times (relative to memory access times) points to larger page sizes. The trend to higher I/O bandwidth points to larger page sizes.

In addition, there are two other important reasons. For a TLB of a given size, the larger the page size, the more address space that is covered by the TLB. For instance, if the page size is 512 bytes, a 64-entry TLB maps only 32k bytes of the address space. If the page size is 4096 bytes, the TLB maps 256k bytes of the address space. The second reason is ease of management of the page tables. Since page tables themselves consume a lot of space, most operating systems need to swap the page tables (e.g., one does not want the page tables associated with an inactive process to consume space in memory). Having 4k-byte pages is optimal for this, because the page tables and page-table directories can also be 4k bytes in size. With smaller pages, either the page tables are larger than a page, requiring added complexity in the operating system to swap them and to allocate multiple contiguous pages to hold them, or one is forced to map virtual addresses via three levels of tables, increasing the time taken to traverse the tables when the page is not in the TLB.

Why provide multiple ways of mapping regions (Figure 4-2)? In many system environments (e.g., real-time operating systems), there are many "tiny" processes. If each process were required to map each of its regions via two levels of tables (directory and page table), the minimum process size would be about 24k bytes if all three regions were used (three directories, three page tables, and three pages), or about 8k bytes if only one region were used.** If we use the option of mapping a region into a single page, the minimum process size is 12k bytes (3 regions), or 4k bytes if only one region is used.

The rationale for the middle mechanism (mapping a region by one page table) is that an operating system will typically use the three regions for code, data, and stack, but although the code size is known in advance, one cannot usually predict the stack size (nor, sometimes, the data size if dynamic memory allocation is supported). Mapping the region used for the stack by one page table gives the oper-

---

*Remember that one can exploit the mapping in a static system for purposes of protection without needing to perform swapping.

**The reason for the sizes 24k and 8k, instead of 36k and 12k, is that the first-level table, in this case the directory, is permitted to be as small as 64 bytes.

ating system the ability to start the program with a small stack and expand it as needed.

Because of the way the TLB is designed in the 80960MC, the four regions associated with one process must have distinct descriptors (i.e., two or more regions cannot have the same descriptor). However, if one wishes, multiple region descriptors can point to the same page table or directory.

ADDRESSING IN THE LARGE

We are now in a position to tie everything together and discuss in more detail how addresses are formed. We will start with Figure 4-3, which shows the relationships among some of the important data structures defined by the architecture. The roots of the relationships are two data structures whose physical addresses are specified at initialization: the processor control block (PRCB) and the segment table. The use of the term "segment table" does not mean that the protected architecture has a segmented addressing structure in the sense of the 80286 and 80386; as discussed earlier, all addressing is done within the confines of a 2^{32}-byte linear address space.

As discussed in Chapter 2, the PRCB contains the physical addresses of the fault and interrupt tables. The PRCB also defines region 3 (of all address spaces) by pointing to an entry in the segment table, which in turn defines region 3. Depending on the mapping selected for region 3 (Figure 4-2), there may be page tables on the path from the segment table to the physical pages that comprise region 3. In other words, the entry in the segment table for region 3 might point to a single page, a page table, or a page-table directory.

Also shown in Figure 4-3 is a process control block (PCB). For simplicity's sake, only one is shown, although many might exist in a typical system environment. The PCB defines the three regions comprising the process's address space by pointing to entries in the segment table, which in turn define the mapping of the regions.

References through the segment table to entities such as regions and PCBs are shown as dashed lines. Such references use a 32-bit value called a *segment selector*, the format of which is shown in Figure 4-4. Segment selectors appear in certain data structures (e.g., a PCB contains three segment selectors for the three regions) and as operands of certain privileged (supervisor-mode) instructions.

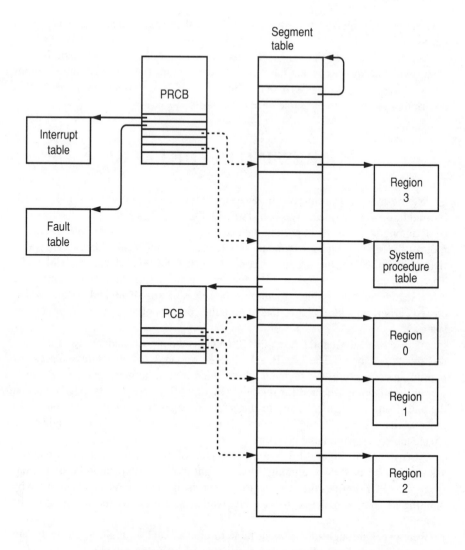

Figure 4-3: Major data-structure relationships

Finally, note that the segment table in Figure 4-3 contains a reference to itself (the arrow pointing back to the top of the segment table). The rationale for this is that, depending on the software environment, the segment table could be large and may need to be dynamically extended, meaning that it would be undesirable to require the segment table to be contiguous in memory. Therefore, the segment table can be a single page or described by a page table (meaning in addition that the entire segment table need not be in memory). The eighth entry in the segment table defines the segment table as a "segment" and can point to one page or to a page table. This means that after one defines the physical address of the start of the segment table at initialization, the processor can access the eighth entry (which always falls in the first or only page of the segment table) and determine how the table is mapped.

Figure 4-4: Segment selector

Entries in the segment table are 16 bytes in length. Except for the eighth entry, the architecture does not specify a particular usage for any entry.* The protected architecture defines ten types of entries, as shown in Figure 4-5.**

All addresses appearing in segment table entries are physical addresses. Except for the embedded entry type, the first two words of each entry are undefined and available for use by software.

The simple-region entry defines a one-page region. The entry points to the start of the page. The V (valid) flag indicates, if set, that the entry is valid; if clear, a reference to the region results in a virtual-memory fault (thus an operating system could use this bit to indicate that the page has been swapped out). The A (accessed) flag is set by the processor when a virtual address to this page is translated. The L (altered) flag is similar, but is set only for a write operation. These three flags also appear in page-table entries.

The C flag, which also appears in other entry types and in page tables, specifies whether the memory reference that is translated via this path is cacheable. This attribute has nothing to do with caching in the 80960MC itself (e.g., instruction cache, TLB), but controls an output pin that may be sampled during the address cycle of the memory reference.

---

*The first seven entries should not be used, however, as they are used by some of Intel's debugger products.

**Again note the presence of "magic" values, in this case for upward compatibility to a higher architecture level. If one encounters a fault whose type or subtype is not defined in the protected architecture level, the likely cause is a violation of a magic value.

Simple region

Page address		000000000000					
	111111			1	C	1	L	A	01	V

Port

Port address		000000		
0101	010000		1 C	111011

Paged region

Page table address	000000		
Size		10	V

Procedure table

Procedure table address		000000	
0011	010000	1 C	111011

Bipaged region

Page table directory address	000000		
Size		11	V

Small segment table

Segment table address	000000000000		
111111		1 C	111011

Embedded

Data	
Type	001

Large segment table

Page table address	000000000000	
111111		101

Process

Process control block address	000000		
0010	010000	1 C	111011

Invalid

	000

Figure 4-5:   Segment table entries

Since this pin is interpreted by the system designer (if at all), the designation "cacheable" is more a matter of intent than definition. Assuming that one uses the pin to control whether an external cache is allowed to hold the memory location, this attribute can denote pages associated with memory-mapped I/O devices or pages containing shared memory in certain types of multiple-processor designs.

Two other entry types define a paged region and a bipaged region. The former points to a page table; the latter to a page-table directory. In both cases, the table or directory must begin on a boundary that is a multiple of 64 bytes, and cannot span a page boundary. This means that a small paged region need not consume an entire page just for the page table.

The *size* field specifies the size of the page table or directory in units of 16 × (size+1). In other words, the size of a page table in bytes is 64, 128, ..., 4096. The *V* flag can be used to indicate that the page table or directory is swapped out.

The next type of entry, the embedded entry, has only one purpose currently defined by the architecture - it contains architecture-defined semaphores. As will be discussed later, the architecture has two basic usage models for process management, which we can name *manual* and *automatic*. The embedded entry is of interest only when the automatic model is used.

The process entry points to a PCB, which must be aligned on a 64-byte boundary and cannot cross a page boundary.

The port entry points to an architecture-defined data structure known as a port. As was the case for an embedded entry, the port entry is of interest only when the automatic process-management model is used.

The next two entries, one of which must be the eighth entry in the segment table, define the mapping of the segment table. A small segment table is one that occupies a single page (256 entries). A large segment table is mapped by a 4k-byte page table, meaning that the segment table can occupy from 1 to 1024 pages, or contain up to 256k entries.

Data structures such as PCBs, ports, and the segment table itself must be capable of being directly referenced by software, typically the operating system. The suggested way to do this is to map the data structures into region 3 of the address space (in addition to mapping them via the segment table).

ADDRESS TRANSLATION

Except for the simple, one-page region, virtual addresses are translated by indexing a page table or a page-table directory and page table. Page tables and page-table directories contain four-byte entries whose format is shown in Figure 4-6.

An entry in a page-table directory either contains the physical address (on a 4k-byte boundary) of a 4k-byte page table or is an invalid entry. Invalid entries are used by software to indicate that the page table is not resident in memory (where the upper 31 bits of the entry might be used to contain the disk address) or to indicate that the part of the address space associated with this entry has not been allocated.

Valid page table directory entry

Page-table address		rt	1

Valid page table entry

Page address		1	C		L	A	rt	1

Invalid entry

	0

BITS

page-table address	12-31	address of page table
rt	1-2	access rights
		00: no user access, supervisor read-only access
		01: no user access, supervisor read/write access
		10: user read-only access, supervisor read/write access
		11: user read/write access, supervisor read/write access
page address	12-31	address of page
C	6	cacheable
L	4	altered
A	3	accessed

Figure 4-6: Page-table entries

An entry in a page table either contains the physical address of the associated page or is an invalid entry. Page-table entries also contain cacheable, altered, and accessed flags as described previously. Typically the accessed flag (which is set by the processor and periodically cleared by the operating system) is used by the operating system to help make decisions about which page should be swapped out of memory, and the altered flag is used to determine whether the page actually needs to be copied out, or whether it can just be overwritten.

The access-rights field in both types of entries is one of the two basic mechanisms provided for memory protection (the other being the ability of different processes to have different address spaces). It allows the operating system and application program to reside in the same address space, providing the means to protect the operating system's code and data from the application program.

In a bipaged region, there are two sets of rights on the address-translation path: those in the page-table directory entry and those in the page-table entry. In this case the *effective rights* are the minimum of the two values.

The two sets of rights are provided in order to enable processes to share parts of the address space. Two processes can share one or more regions by having the appropriate segment selectors in the PCBs index the same entry in the segment table, but this approach may not be sufficiently granular in some cases, and it gives both processes the same rights to the shared address space.

As an example, suppose we wish to share four megabytes of process A's space with process B and that we want process A to have read/write access to this space but to restrict process B to read-only access. Process A would map this space into its address space with a page table and page-table directory, and all the entries in the page table, and the corresponding entry in the directory, would be encoded with read/write access. One of B's page-table directory entries would point to the same page table (i.e., the page table is shared), but this entry would be encoded with read-only access.

One might also attempt to share parts of the address space by using separate page tables and having their corresponding entries point to the same addresses. This technique will work, but is not advisable if one is implementing paging (page swapping) because multiple page-table entries point to the same page. This means that one must look at all of these entries when examining the accessed/altered status of the page and update all of these entries when swapping the page into or out of memory.*

Note that there are no rights bits present when a region is a simple region (a single page). In this situation the rights are implicitly read/write for both user and supervisor.

The actual address-translation steps are summarized below. Note that this discussion is only of the logical steps performed; the existence of the TLB in the processor allows almost all memory references to be translated directly in a single cycle without referencing the tables in memory.

Bits 30-31 of the address tell us which of the four current regions is being accessed. If the region is a simple region, bits 12-29 of the virtual address must be zero (otherwise a fault occurs). The physical address in the page-address field in the region entry is concatenated to bits 0-11 from the virtual address.

If the region is a paged region, the value represented by bits 16-29 of the virtual address must not be greater than *size* (a field in the region entry); otherwise a fault occurs. The value of bits 12-21 of the virtual address indexes an entry in the page table. The physical address in the page-address field in the page-table entry is concatenated to bits 0-11 from the virtual address.

---

*The latter part of this also applies when page tables are shared and one is swapping the page tables.

If the region is a bipaged region, the value represented by bits 26-29 of the virtual address must not be greater than the *size* field. Bits 22-29 index an entry in the page-table directory. Bits 12-21 index an entry in the associated page table, and the physical address is again the page-address from this entry concatenated to the low-order 12 bits of the virtual address.

During address translation a number of faults can occur. The faults are summarized below.

Invalid STE   The segment-table entry is the invalid entry.

Invalid region descriptor   The V flag in the entry is clear.

Invalid directory entry   The low-order bit of the page-table directory entry is clear.

Invalid page-table entry   The low-order bit of the page-table entry is clear.

Length   The virtual address points beyond the defined size of the region.

Rights   The access is not permitted because of insufficient rights.

In the event that the segment table itself is paged, a similar address-translation process occurs to locate the entry corresponding to the index in the segment selector.

MANUAL PROCESS MANAGEMENT

The protected architecture provides two usage models of process management: *manual*, where the operating system is entirely in charge, and *automatic*, where low-level process-management functions are performed directly by the processor, and where there is added support for managing processes in a multiple-processor configuration. Since the manual model is much simpler than the automatic model, we will start with it.

The architecture defines processes with the PCB data structure; the parts of the PCB that apply when using manual process management are shown in Figure 4-7.

Except for some timing information and the resumption-record field, the PCB contains only the registers associated with the process (process, trace, and arithmetic controls, region segment selectors, and global registers). When a process is not the current process, the values of its registers are stored in the PCB.*

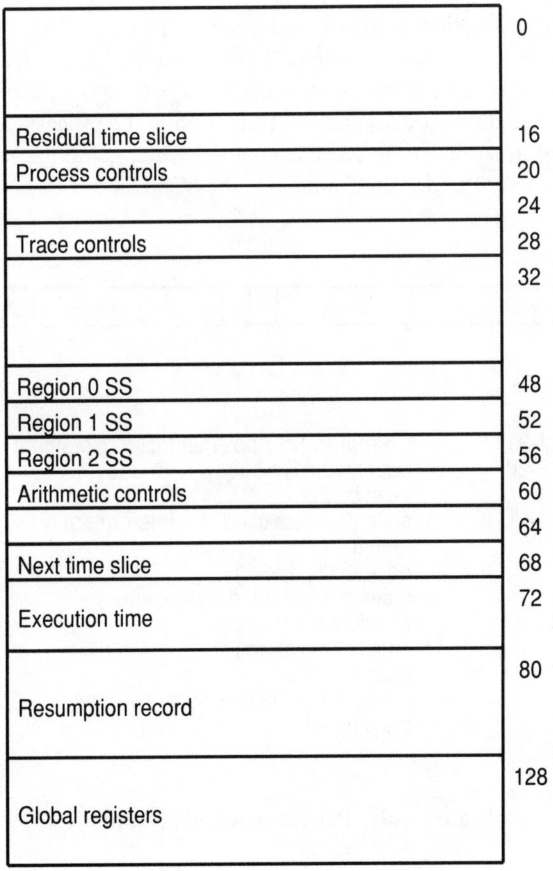

	0
Residual time slice	16
Process controls	20
	24
Trace controls	28
	32
Region 0 SS	48
Region 1 SS	52
Region 2 SS	56
Arithmetic controls	60
	64
Next time slice	68
Execution time	72
Resumption record	80
Global registers	128

Figure 4-7:   Process control block (manual)

The RESUMPRCS (resume process) instruction has one operand, a segment selector to a PCB. RESUMPRCS loads the information from the PCB into the

---

*The local registers are not in the PCB because they are stored in the process's stack frames.

processor, thus changing the execution environment to that of this process. The SAVEPRCS (save process) instruction performs a "checkpoint" of the current process; the state of the process is copied into its PCB in memory, the local-register sets are copied into their associated stack frames in memory, and execution continues in this process.

Note that the instruction pointer is not contained in the PCB. The RESUMPRCS instruction obtains the IP from register R2 of the current frame (registers R0-R15 are loaded from the first 16 words of the current stack frame).

The protected architecture also adds information to the process-controls register. The pertinent information in the process controls when using the manual model is shown in Figure 4-8. The new information is the *rf*, *tm*, and *ts* flags.

internal state	priority	--	0	st	rf	--	tfp	res	tm	ts	0	---	em	t

BITS

internal state	21-31	internal state - implementation dependent
priority	16-20	priority (0-31)
0	14	must be zero
st	13	state (0 = executing, 1 = interrupted)
rf	12	refault
tfp	10	trace fault pending
res	9	resume suspended instruction
tm	8	enable timing
ts	7	enable time slicing
0	6	must be zero
em	1	execution mode (0 = user, 1 = supervisor)
t	0	trace enable

Figure 4-8:   Process controls (manual)

Two of the added flags, and additional information in the PCB, are associated with timing functions. The processor maintains the elapsed execution time of each process and can interrupt a process when a time limit expires, thereby removing some of this burden from the operating system and the surrounding hardware system. Time is maintained in units known as *ticks*, which the architecture permits to be implementation dependent. In the 80960MC implementation, a tick is 8 microseconds at 16 MHz operation (32 MHz external clock) and 6.4 microseconds at 20 MHz operation.

The processor contains a downward-counting timer register into which the residual-time-slice field from the PCB is loaded when the RESUMPRCS instruction is executed. The timer is stored in this field when the SAVEPRCS instruction is executed. If the *tm* flag is set, the timer is decremented every tick. If, at any point in time, the timer is zero and *tm* is set, the processor loads the timer from the next-time-slice field in the PCB, increments the value of the 64-bit execution-time field in the PCB by this value, and tests the *ts* flag. If *ts* is clear, execution continues; if *ts* is set, a time-slice fault occurs.*

The effect of the above is that one has three timing options: no timing, timing with no time slicing, and time slicing. The latter is typically used in multi-user, timesharing systems to ensure that no process "hogs" the processor. Let's say, for purposes of discussion, that the operating system wishes to give all processes a time slice of 100 milliseconds. This value (in ticks) is placed in the next-time-slice field in the PCBs, and also in the residual-time-slice field when the process is created. When execution of a process begins, it executes for 100 milliseconds and then encounters a time-slice fault, at which time the operating system can obtain control and perhaps switch to another process. Alternatively, the process may be suspended prior to the completion of its time slice, for example if an interrupt is received and the operating system decides that another process should execute at this point, or if, within a call by the process to the operating system, the operating system decides that the process cannot continue execution at this time (e.g., the process needs to wait for the completion of an I/O operation).

When the process is suspended at other than the end of a time slice, the current value of the timer is stored in the residual-time-slice field in the PCB. For example, if the process had executed 70 milliseconds of its time slice and was then suspended, the processor would reload the timer with the remaining 30 milliseconds when the process is resumed (unless the values are changed by the operating system in the meantime).

A process can obtain its elapsed execution time by executing the LDTIME instruction. This instruction reads the execution-time field from the PCB, subtracts the current value of the timer from it, and puts the 64-bit result in the destination register pair.

Finally, the resumption record in the PCB is used as a temporary storage area for instruction resumption. In certain situations, the RETURN instruction (e.g., associated with an interrupt that caused an instruction to be suspended) copies resumption information from the stack into this area of the PCB. When the *res* flag is set at the start of execution of an instruction, the processor obtains the resumption information from the PCB.

*The 80960MC (as opposed to a different implementation of the protected architecture) uses only the low-order 16 bits of the two time-slice fields in the PCB and ignores the upper 16 bits. Thus, at 20 MHz, the maximum time interval is about 0.4 seconds. Also, one should ensure that the next-time-slice field is set to 16 ticks or greater; if not, the processor can get stuck in an endless loop (because the timer can expire before the processor has restarted execution or invoked the time-slice fault handler).

AUTOMATIC PROCESS MANAGEMENT

The automatic process-management usage model extends the manual model by allowing the processor to perform many of the low-level, time-critical, functions of an operating system. The processor automatically dispatches processes from a queue of ready processes and reschedules the processes back onto the queue, without any software intervention. The processor also provides two interprocess communication mechanisms (*messages* and *semaphores*), where the scheduling for execution of processes waiting for messages or semaphore signals is again done without software intervention. Finally, these functions are designed for correct behavior in the presence of multiple processors.

The central data structure in the automatic model is the *dispatch port*, which is a queue of ready processes (i.e., processes that are waiting to be executed). Since the PCB contains an image of the process-controls register, and since one of the fields in the process controls is priority, each ready process has an associated priority (0-31). The dispatch port actually contains 32 queues, one for each priority level. Thus a ready process of priority i would be enqueued in queue i in the dispatch port.

The relationships among the processor, the dispatch port, and ready processes are shown in Figure 4-9. The dispatch port points to 32 queues of PCBs, some or all of which may be empty. Here we have shown two queues of two PCBs each. When the processor is idle, it locates the highest priority nonempty queue, removes the first PCB from the queue, and resumes execution of this process (i.e., does something similar to what the RESUMPRCS instruction in the manual model does). If the dispatch port happens to be empty, the processor rechecks it at periodic short intervals.

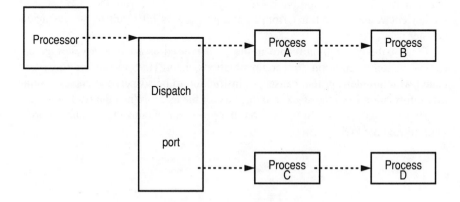

Figure 4-9: The queue of ready processes

After a process is dispatched, the processor continues to execute it until one of several events occurs. For example, the process may block itself (i.e., go off to wait on a message queue or a semaphore). In such a situation, the processor suspends the process (i.e., does something similar to the function of the SAVEPRCS instruction) and goes back to the dispatch port to find another process to execute. Another situation is that where the process's timer hits zero (i.e., the time slice has expired). In this case, one has two options (determined by a flag in the process's process controls): one can have the processor suspend the process and reschedule it by placing it at the end of the queue corresponding to the process's current priority, or one can have a fault occur, allowing the operating system to determine what to do with the process. The operating system may wish to adjust the process's priority (to allow lower priority processes to execute) before enqueuing the process back on the dispatch port.

The relationships in Figure 4-9 are shown as dashed lines to indicate that they are segment selectors. Also, the figure shows just one processor pointing to the dispatch port. However, in multiple processor systems, one has many options. All processors could point to one dispatch port, in which case processes hop from processor to processor in a load-balancing fashion. Conversely, each processor could have its own dispatch port. Or, one could configure the system as a mixture of the two, that is, some, but not all, processors could share a dispatch port.

The structure of a dispatch port is shown in Figure 4-10. For each priority queue, the dispatch port contains a segment selector (SS) to the first and last PCBs in the queue. The queue-status field is a bit vector, such that if bit i is set, the priority i queue is not empty. This allows the processor to find the highest priority nonempty queue quickly by finding the most significant set bit.

qs ty	Lock	0
Queue status		4
Priority 0 queue head SS		8
Priority 0 queue tail SS		12
Priority 31 queue head SS		256
Priority 31 queue tail SS		260

Figure 4-10:   Dispatch port

The *lock* field permits multiple processors to use the same dispatch port without tripping over themselves. If the low-order bit of the lock is set, the dispatch port is said to be locked. When the processor needs to manipulate the port, it sets the lock (by the same algorithm as in the ATMOD instruction) and examines the previous value of the lock. If the lock was not already set, the processor proceeds with its operation on the port; otherwise, it repeats the attempt to get the lock.* The lock can also be used by software (by using the ATMOD instruction) if the operating system needs to manipulate the port itself.

The remaining two bits in the port will be revisited later, since the dispatch port is actually a specific use of a more-general data structure known as a *port*. In a dispatch port, *qs* would be 0 and *ty* would be 1.

When using the automatic model, several additional fields of the PCB are used; this view of the PCB is shown in Figure 4-11. The first two words are used for queuing the PCB. When the PCB is in queue (e.g., on a dispatch port), the root SS points back to the "root" of the queue (e.g., to the port) and the link SS points to the next element of the queue. The received-message field is used as a temporary storage area by the message communication functions, as is the next field (an SS to a dispatch port).

The lock field in the PCB is similar to the lock in the port. When the processor dispatches a process, it sets its lock, which remains set until the process is suspended. Since, in the automatic model, the operating system is not directly involved in the dispatching of processes, the operating system can examine the lock in an arbitrary PCB (in a multiple processor system) to determine whether the process is currently being executed.

The last additional field, the *process notice*, solves a subtle problem in multiple processor systems, namely the "chasing down" of a process. After a process is dispatched, but before the first instruction is executed, bits 16 and 31 of this word in the PCB are checked.** If they are both set, they are cleared and an *event* fault occurs. This allows an operating system to take control of a process in the set of ready processes without the race conditions that would otherwise occur. However, since the check is made only during the dispatch step, it doesn't solve the problem if the process is already in execution. A special IAC message exists to solve this part of the problem.

---

*In the 80960MC, the processor will make this attempt approximately once every tick, or about every 6.4 microseconds at 20 MHz.

**The use of two bits instead of one allows software to use one as an "enable" flag and the other as a "request" flag.

Link SS		0
Root SS		4
Received message		8
Dispatch port SS		12
Residual time slice		16
Process controls		20
Process notice	Lock	24
Trace controls		28
		32
Region 0 SS		48
Region 1 SS		52
Region 2 SS		56
Arithmetic controls		60
		64
Next time slice		68
Execution time		72
Resumption record		80
Global registers		128

Figure 4-11: Process control block (automatic)

When using the automatic process model, there are additional flags in the process controls, as shown in Figure 4-12. Flag *tsr*, if set, directs the processor to reschedule the process itself; otherwise, the time-slice fault occurs. Flag *pr*, if set, denotes a *preempting* process. This is associated with how processes are treated when they become unblocked as a result of being sent a message or a signal.

internal state	priority	--	0	st	rf	pr	tfp	res	tm	ts	tsr	---	em	t

BITS

	BITS	
internal state	21-31	internal state - implementation dependent
priority	16-20	priority (0-31)
0	14	must be zero
st	13	state (0 = executing, 1 = interrupted)
rf	12	refault
pr	11	preempt
tfp	10	trace fault pending
res	9	resume suspended instruction
tm	8	enable timing
ts	7	enable time slicing
tsr	6	time-slice reschedule
em	1	execution mode (0 = user, 1 = supervisor)
t	0	trace enable

Figure 4-12:   Process controls (automatic)

Message Passing

The dispatch port is a special case of the more general function of message passing. One can view the dispatch port as a message queue, in which the messages are processes ready for execution and the recipient of a message is a processor. In the general case, messages are data structures (which could be, for instance, PCBs or "just plain data") and the recipients are processes. Because segment selectors (SSs) are used to link the queues, messages must be one of the entities that are pointed to by segment-table entries (Figure 4-3). The first two words of the message are used for linkage in the same way as the first two words of a PCB (Figure 4-11).*

The root of a message queue is called a *port*, and it has the same structure as a dispatch port (Figure 4-10). A message queue can be in one of the following mutually exclusive states:

- Empty (queue-status bit vector is zero)
- Contains messages (bit vector is nonzero and flag *qs* is clear)

---

*A common trick is to use a "fake" three-word PCB as a message.

- Contains waiting (blocked) processes (bit vector is nonzero and flag qs is set)

The architecture defines two varieties of ports. If flag ty is 1, the port contains 32 priority queues (a *priority* port); otherwise, the port data structure contains just three words, where the first is the same as for the dispatch (priority) port and the second two are the head and tail of a single queue. In such a port, known as a *FIFO* (first-in/first-out) port, the priorities of messages and waiting processes are ignored.

The SEND instruction, a privileged instruction, has three operands: segment selectors to a port and a message, and a priority. If the port is empty or contains messages, the message is enqueued at the port (at the end of the specified priority queue, for a priority port, or at the end of the single queue, for a FIFO port). If, however, the port contains waiting processes, the first highest-priority PCB is dequeued, the message SS placed in the PCB (which will eventually be returned as the result of the process's suspended receive instruction when the process is dispatched), and the process "unblocked" by enqueuing the PCB at the front of the proper priority queue in the dispatch port pointed to by the PCB.

The RECEIVE instruction, another privileged instruction, has one input operand (SS of a port) and returns the SS of a message. If the port is empty or contains waiting processes, the current PCB is enqueued at the end of the appropriate priority queue of the port (or on the only queue, in the case of a FIFO port). The instruction pointer is not advanced, and the *resume* flag is set in the process controls so that the processor remembers to have the reexecution of the RECEIVE instruction obtain the message from the third word in the PCB if and when the process resumes execution. If the port contains messages, the first of the highest priority messages is dequeued and its SS is returned as the result of the receive instruction.

Another form of the RECEIVE instruction, CONDREC (conditional receive) is provided for processes that do not wish to wait if no message is available. CONDREC returns a message if one is available and sets the condition code to indicate whether a message was dequeued.

Whenever one of these instructions is executed, the port is locked in the same way that the processor locks the dispatch port when manipulating it. This ensures correct operation in multiple processor systems, where it is possible for a process on one processor to be sending a message to a port at the same time that a process on another processor is executing a RECEIVE instruction on the same port.

In the discussion of the SEND instruction we mentioned that if a process is unblocked, it is dequeued from the port and enqueued on the dispatch port. One added complexity can be introduced now - the concept of *preempting processes*. After the unblocked process is enqueued on the dispatch port, the *pr* flag in the process controls of that process is checked. If *pr* is set, if the current process is not in the interrupted state, and if the priority of the unblocked process is greater than that of the current process, the processor suspends the current process, enqueues it at the front of the appropriate priority queue on the dispatch port, and dispatches the highest priority process on the dispatch port. Thus, if *pr* is set in a process, it tells the processor that if that process is ever unblocked by a lower priority process, the processor should abandon the lower priority process.*

Semaphores

The architecture defines a second interprocess communication mechanism - *counting semaphores*. The use of a semaphore can be thought of as a special kind of message, where the information being communicated is not a message, but the fact that a message has been sent. Thus semaphores are used to communicate signals rather than messages.

The data structure analogous to the port is the semaphore. The semaphore is an unusual data structure in that it appears *inside* a segment-table entry, rather than being pointed to by one. The entry used is the embedded entry (Figure 4-5).** The format of the semaphore is shown in Figure 4-13.

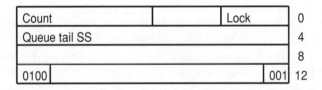

Figure 4-13: Semaphore in an embedded segment-table entry

Like a port, a semaphore can be in one of three states: it can be empty, contain signals (count field is nonzero), or contain waiting processes (count is zero and queue root is nonzero). There is only a single process queue, which is maintained in priority order. The semaphore points to the last process in the queue, and the link field of the last process points to the first process in the queue (in other words, the list is circular).

The SIGNAL instruction is analogous to the SEND instruction. It has one operand: a segment selector to a semaphore. If the semaphore is empty or contains signals, the count is incremented. If the semaphore contains waiting processes, the first

---

*There is another preemption option pertaining to multiple processor systems; it is discussed later.

**The rationale for this is to save memory space. A semaphore is only two words long, but if it were pointed to by the entry, rather than being contained within the entry, it would consume a total of 80 bytes of space rather than 16. The reason that it would consume 80 bytes is that entities pointed to by segment-table entries must be aligned to 64-byte boundaries.

process in the queue is unblocked in the same way as in the SEND instruction; the possibility of preemption is also present.

The WAIT instruction is to semaphores what RECEIVE is to ports. If the semaphore is empty or contains other waiting processes, the current PCB is placed in the list, which is ordered by priority, and ordered FIFO within priority); the process is suspended; and the processor attempts to dispatch another process. As is not the case with the RECEIVE instruction, the instruction pointer *is* advanced, since the WAIT instruction has no result operand. If the semaphore contains signals, the count is decremented.

There is also a CONDWAIT (conditional wait) instruction, which never causes process suspension. It sets the condition code to indicate whether the count field was decremented.

Like a port, a semaphore contains a lock, which is used by the processor when manipulating the semaphore to ensure correct operation in multiple processor systems.

PROCESSOR CONTROL BLOCK

In Chapter 2, the PRCB was introduced in the core architecture as a data structure out of which the processor obtained a few key addresses at initialization, such as the addresses of the fault table, interrupt table, and interrupt stack. In the protected architecture, particularly when the automatic process-management model is used, the PRCB takes on additional significance.

The content of the PRCB is shown in Figure 4-14. The current-process, dispatch-port, resumption-record, and idle-time fields are usually of significance only when the automatic process model is used. The current-process field holds the address of the currently executing process. In a multiple processor system, having information about which process is currently running on each processor is sometimes useful to an operating system in sending certain types of interagent communication (IAC) messages. The dispatch-port field identifies the dispatch port from which the processor should dispatch processes. The resumption record serves the same purpose as the resumption record in the PCB. It exists in the PRCB because there are situations in which the processor is executing instructions that might be suspended when the processor has no current process.*

The idle-time field contains a 64-bit value (in units of ticks) that the processor increments when it is in a state known as *idle*. When in this idle state, the processor does not increment the field in memory at every tick boundary, but updates the value at larger periodic intervals; therefore it is only an approximation of the time

---

*An example is when an interrupt occurs when the processor is idling because there are no ready processes, and then a second interrupt causes an instruction in the interrupt handler to be suspended.

that the processor is idle. The main use of the idle-time value is for software load
measurement in a multiple processor system.

Implementation dependent	0
Processor controls	4
	8
Current process SS	12
Dispatch port SS	16
Interrupt table physical address	20
Interrupt stack pointer	24
	28
Region 3 SS	32
System procedure table SS	36
Fault table physical address	40
Implementation dependent	44
	48
Implementation dependent	
	64
Idle time	68
System error fault	72
	76
	80
Resumption record	
	128
System error fault record	

Figure 4-14: Processor control block

When address translation is enabled, the fault and interrupt tables are located by
physical (untranslated) addresses. However, the interrupt stack is located by virtual
address so that one does not have to allocate contiguous physical pages for it.

The PRCB contains three fields defined as implementation dependent. The first two are associated with debugging features of the chip that Intel defines as proprietary. These fields should be initialized to zero and never changed. The third implementation-dependent field is associated with process preemption in multiple processor systems, which is discussed later.

The processor contains another control register, the *processor-controls* register, which is loaded from the PRCB during initialization. No instruction is provided to examine or modify this register; instead, a special IAC message is provided for doing so. The register is shown in Figure 4-15.

wep	--	internal state	interim priority	--	cdp	adt	nonpreempt limit	0	state	mpp	0

                    BITS

	BITS	
wep	31	implementation dependent
internal state	21-26	implementation dependent
interim priority	16-20	implementation dependent
cdp	11	check dispatch port
adt	10	enable address translation
nonpreempt limit	5-9	priority boundary for preemption
state	2-3	processor state
		00: stopped or stopped-interrupted
		10: idle or idle-interrupted
		11: executing a process
mpp	1	enable multiple processor preemption

Figure 4-15: Processor controls

The important flag for enabling address translation (*adt*) is located in the processor controls. Another important field is the processor state. When combined with the state flag from the process controls, we see that the processor is always in one of the following five states:

stopped   The processor is doing nothing. Its priority is zero. The only way of escaping from this state is via an IAC message.

idle   Similar to the stopped state, except the processor periodically checks the dispatch port. If it finds a ready process, it will enter the executing-a-process state.

idle-interrupted   The processor is executing an interrupt han-
                                    dler. When the last interrupt handler returns,
                                    the processor will enter the idle state.

executing a process   The processor is executing a process.

executing a process, but interrupted   The processor is executing an interrupt han-
                                    dler. When the last interrupt handler returns,
                                    the processor will enter the executing-a-pro-
                                    cess state.

The *cdp* flag is an indicator that the processor uses in certain preemption situations; it is not a flag that software normally uses, and it should be initialized to
zero.* When unblocking a process, the processor does not attempt to perform process preemption if the current process is in the interrupted state.** Instead, the processor sets the *cdp* flag. During a return from an interrupt handler in the situation
where the processor state is changing from "executing but interrupted" to
"executing," *cdp* is checked. If set, it is cleared and the processor looks at the dispatch port to see if a higher priority process is ready. Thus, *cdp* is essentially a way
for the processor to remind itself to check for a process preemption at a later point
in time.

Other information in the processor controls is associated with preemption and is
discussed later in this chapter.

A Note on Segment Selectors

As shown in Figure 4-4, whenever a program stores an SS in an architecture-
defined data structure or uses an SS as an operand of an instruction, the low-order
six bits must be set. Failure to do so could result in a fault.

However, there are cases where the *processor* stores an SS, either implicitly or
as an indirect action. Examples are the current-process SS in the PRCB and the
linkage SSs in ports, semaphores, and messages. These SSs do not necessarily have
their low-order six bits set, so they are more properly regarded as segment indexes
rather than complete SSs. If software fetches one of these for use as an SS, it
should first set the low-order six bits.

---

*Initializing it to zero is particularly important when using the manual process model, where there is no
dispatch port.

**The reason is that when a process is interrupted, the interrupt handler essentially "piggybacks" on the
process. Having the processor suspend the process during this time would cause the interrupt handler to
disappear, possibly cause the interrupt stack to be overwritten, and in short lead to general mayhem.

DATA TYPES

The protected architecture does not have any additional data types, but it does make one key change in the addressing of operands. Operands still must be aligned on their natural boundaries in registers (e.g., a 64-bit operand must be in an even/odd register pair), but they need not be aligned on natural boundaries in memory. For instance, the LD instruction may specify any byte boundary in memory, and this byte and the next three bytes are loaded into the designated register.

Why is this the case in the protected architecture when we argued for natural boundary alignment in Chapter 2? The first reason is that the 80960MC will be used in graphics applications, where long bit strings that start and end on arbitrary bit boundaries are frequently moved and manipulated. Such algorithms can be written using the core architecture (with natural boundary alignment), but they prove to be significantly slower than if words can be addressed on any byte boundary. The second reason is that an implementation of the protected architecture is significantly more complex than one of the core architecture, and thus the added complexity of unaligned operands is not as painful.

It is important to realize, however, that since accesses to unaligned operands in the 80960MC are usually slower than accesses to aligned operands, one should continue to align operands wherever possible. Also, because of the protocol of the 80960's bus, operands that span 16-byte boundaries result in even more overhead.

INSTRUCTION SET

The protected architecture adds 17 instructions to those of the numerics architecture. All of the additional instructions, which are listed in Table 4-1, have the REG format. In addition, a few of the core instructions have expanded semantics in the protected architecture.

The string instructions are provided at this level (rather than at the core or numerics level) because other aspects of the protected architecture (e.g., process management) require a microcoded implementation of the architecture, as do the string instructions. Since compilers must implement string operations in the core and numerics levels using loads and stores, one might ask why the string instructions are provided at all. There are three reasons:

1. To produce optimal performance, whether implemented in software or microcode, string algorithms need to take into account characteristics of the processor's bus and internal organization. Therefore, a software-implemented string algorithm that is optimized for one implementation of the architecture may not be optimal for another implementation. Since microcode is intimately tied to the implementation, providing the algo-

rithms in microcode hides the algorithms from software and allows each implementation to contain the optimal algorithm.

Table 4-1:   Added instructions

Mnemonic	Function	Sets CCode
movstr	move string	
movqstr	move quick string	
fill	fill string	
cmpstr	compare string	X
ldphy	load physical address	
inspacc	inspect access	
send	send message	
receive	receive message	
condrec	conditional receive	X
signal	signal semaphore	
wait	wait for signal	
condwait	conditional wait	X
schedprcs	schedule process	
sendserv	send current process	
ldtime	load process execution time	
saveprcs	save process	
resumprcs	resume process	

2. Because of things like the burst nature of the 80960's bus and its register scoreboard, an optimal software string algorithm would need to use a lot of registers (about 11), which could necessitate the compiler's spilling registers into memory. By implementing the algorithms in microcode, internal registers can be used.

3. Implementing the algorithms in software increases the size of the program and thus pushes other parts of the program out of the instruction cache.

The MOVSTR instruction has three operands: the addresses of the first bytes of the source and destination strings, and the number of bytes to be moved. It copies this number of bytes from the source string into the destination string. It copies the string properly even if the strings overlap; the logical effect is as if the source string is copied into a temporary string, which is then copied into the destination. The

MOVQSTR instruction is similar, except it assumes that the source and destination do not overlap. MOVQSTR is the faster alternative when one knows in advance that the strings do not overlap.

The FILL instruction also has three operands: the address of the first byte of a string, its length, and a 32-bit value. The string is "filled" with the value. FILL would be used, for instance, if one wanted to store "blanks" in every byte of a string. The CMPSTR instruction does a byte-by-byte comparison of two strings and sets the condition code to indicate whether the value of the first string is less than, equal to, or greater than that of the second.

The LDPHY instruction, given a virtual address as its input operand, returns the corresponding physical address as its result. In other words, it performs an address translation and returns the result. One important use for it is in obtaining the physical address of an I/O buffer in order to initialize a DMA channel.

The INSPACC (inspect access) instruction, given an address as its input operand, returns the effective rights on this address path. Its primary purpose is to give an operating system an easy way to check the validity of addresses passed to it from an application program. For instance, suppose we call the operating system to perform an I/O read, but the address we pass it (either unintentionally or maliciously) is one that maps to a page owned by the operating system, a page with only supervisor access rights. If the operating system blindly used the address, any number of undesirable things could happen. However, the INSPACC instruction allows the operating system to check the rights associated with any address parameter it receives, thereby allowing it to take some action if the address is not one belonging to the application program.

The SCHEDPRCS and SENDSERV instructions are specialized instructions associated with the automatic process-management model. The former contains one operand: a segment selector of a PCB. It performs the equivalent of the unblock operation discussed previously in the context of the SEND and SIGNAL instructions, and preemption. It enqueues the PCB at the front of the appropriate priority queue in the dispatch port, and performs the preemption action if so indicated in the PCB. An operating system can make a process a ready process by using the SEND instruction, sending the process as a message to the dispatch port, but the SCHEDPRCS instruction allows the operating system the possibility of having the process dispatched immediately.

The SENDSERV instruction has one operand: a segment selector of a port. It suspends the current process and sends it to the designated port, which causes the processor to dispatch another process. For instance, if the operating system handles ends of time slices by a fault (rather than by having the processor automatically enqueue the process back on the dispatch port), the fault handler could use this instruction to put the process on the dispatch port after adjusting the process's priority.

The significance of supervisor mode in the protected architecture is twofold: all instructions that reference memory when the rights in the page table(s) specify no user access require that the process be in supervisor mode, and certain instructions require that the process be in supervisor mode. These privileged instructions are those that have a segment selector as an operand, namely SEND, RECEIVE, CONDREC, SIGNAL, WAIT, CONDWAIT, SCHEDPRCS, SENDSERV, and RESUMPRCS. In addition, as in the core architecture, one must be in supervisor mode to execute MODPC with a nonzero mask.

INTERRUPTS

Interrupts in the protected architecture have only a few subtle differences from those in the core and numerics levels. One is that two additional flags, *tm* and *ts*, are cleared in the process controls before calling the interrupt handler. This ensures that the current process (if there is one) doesn't "timeout" during the execution of the interrupt handler, since, as discussed earlier, this would prove to be disastrous. It also suspends execution timing of the process so that the execution time of the interrupt handler doesn't get charged to the process.

Another difference when using the automatic process-management model is that interrupts can occur when in the idle state. In the protected architecture, value 110 in the return-status field in register r0 is defined as an interrupt call while in the idle state, and causes the corresponding RETURN instruction to place the processor back in the idle state if there are no pending interrupts.

Since the protected architecture adds the concept of the process, and since certain instructions can refer to attributes of the process (e.g., LDTIME), it is conceivable that one could execute such an instruction in an interrupt handler. Such a practice is undesirable, since interrupts are normally asynchronous and thus the interrupt handler cannot rely on interrupting a specific process. In addition, the situation exists where there is no current process (the interrupt occurs in the idle state). In this situation, the state of the processor architecture is defined only by the state established by the call to the interrupt handler.

There is one situation where the processor makes an implicit reference to the current PCB - when accessing the resumption record for resumption of a suspended instruction. When there is no current process, the processor uses the resumption-record field in the PRCB.

Interrupt handlers must be careful not to do anything that would cause the underlying process to be suspended and rescheduled. Thus, unless one understands precisely the consequences, one should not execute a SENDSERV, WAIT, or RECEIVE instruction, or enable timing, in an interrupt handler.

Because most of the instructions that are added at the protected architecture level can have long execution times, these instructions are treated as resumable instructions in the same way as some of the floating-point instructions, as described in Chapter 3. For instance, if an interrupt occurs during execution of a MOVSTR instruction, the instruction is suspended and sufficient information is stored in the resumption record to enable the instruction to continue from the point of the interrupt when it is resumed.

FAULTS

Because of the presence of address translation, the fault mechanism in the protected architecture is extended in two major ways. First, for example, a virtual-memory fault because of an invalid page-table entry could occur in the midst of a MOVSTR instruction. The instruction cannot simply be aborted and reexecuted later for several reasons, one of them being that the instruction may have already made irreversible changes to the program's state, as, for example, in the case of overlapping strings. This situation is handled by having certain types of faults function in the same way as interrupts, in the sense that the instruction is suspended and a resumption record is stored on the stack with the fault record.

The second extension is to give software the means to recover from faults that occur during the invocation of fault handlers. For instance, suppose we specify in the fault table that floating-point faults are to be handled by a call to a local procedure; that is, the fault record is placed on the current stack, and the fault handler also uses this stack. The storing of the fault record on the stack could result in a virtual-memory fault if, for example, this causes the stack to extend onto a page that is not currently in memory. Since this situation is one that needs to be handled in systems that implement paging, the architecture must accommodate it and satisfy two objectives: (1) the virtual-memory fault must be reported to software and (2) the floating-point fault must not be lost, so that it can be reported again once the virtual-memory fault condition is removed.

The fault record in the protected architecture is expanded to 48 bytes, as shown in Figure 4-16. When a fault handler is called, the fault record is placed on the stack in the 48 bytes immediately preceding the frame for the fault handler. In the case of a fault that causes the faulting instruction to be suspended, a resumption record of 0-48 bytes is placed on the stack preceding the fault record. Thus, when a fault handler is called, 48 + M is first added to the stack pointer, where M is the size of the resumption record.

The information added in the fault handler is associated with *override*, which is the condition where a second fault "overrides" a first fault. The override information consists of a second set of fault information (type, subtype, flags, and fault data).

reserved				0
Override Fault Data				4
Fault Data				16
OFlags	OType		OSubtype	28
Process Controls				32
Arithmetic Controls				36
FFlags	FType		FSubtype	40
Address of Faulting Instruction				44

Figure 4-16: Fault record

Since new fault types are added, the fault table is expanded, as shown in Figure 4-17. Also note that in the core architecture, an entry in the fault table could specify a local procedure or a call via the system procedure table. In the protected architecture, the latter is generalized to allow a call via any procedure table, which is done similar to the way the CALLS instruction specifies a call through the system procedure table (i.e., the procedure table referenced in the PRCB).*

The fault types and subtypes added in the protected architecture are listed in Table 4-2.

---

*This represents the only usage of multiple procedure tables in the protected architecture, and, as such, does not have much obvious added utility. This facility exists for upward compatibility to the extended architecture, which makes much more use of procedure tables.

Override entry	0
Trace fault entry	8
Operation fault entry	16
Arithmetic fault entry	24
Floating-point fault entry	32
Constraint fault entry	40
Virtual-memory fault entry	48
Protection fault entry	56
Machine fault entry	64
Structural fault entry	72
Type fault entry	80
reserved	88
Process fault entry	96
Descriptor fault entry	104
Event fault entry	112

Local Procedure Fault-Table Entry

Procedure address	00
reserved	

Procedure-Table Fault-Table Entry

Procedure number	10
Procedure table SS	

Figure 4-17: Fault table

The virtual-memory fault occurs during address translation if the low-order bit (*V* bit) in the region descriptor, page-table directory entry, or page-table entry is 0. When paging is implemented, this bit is typically used to denote that the directory, page table, or page does not reside in memory (or has not yet been allocated).

If the fault occurred because of an invalid directory entry or page-table entry for a region, the first word of the fault-data field in the fault record contains the virtual address (although its low-order 12 bits are undefined). If the fault occurred because of an invalid region descriptor or because of an invalid page-table entry for a large segment table (meaning that the needed part of the segment table is not in memory), bits 5-31 of the second word of the fault-data field contain the segment index. Virtual-memory faults can result in instruction suspension.

The protection fault indicates an attempted protection violation. The length subtype indicates that the violation was (1) referencing a procedure number greater than 259 in a procedure table, (2) using an address that cannot be translated because it falls beyond the defined extent of a directory or page table, or (3) using a segment index that is beyond that of the last entry in the segment table. The rights subtype indicates that the violation occurred because the effective rights on the addressing path were insufficient; in this case bit 0 of the *FFlags* field in the fault record is used to denote whether the access was a read (0) or write (1). In the fault-data field, the first word is the virtual address (with the 2 high-order bits and 12 low-order bits undefined) and bits 5-31 of the second word contain the segment index. As with virtual-memory faults, protection faults can result in instruction suspension.*

Table 4-2: Added fault types and subtypes

Type Encoding	Subtype Encoding	Description
6		Virtual-memory fault
6	1	Invalid region descriptor
6	2	Invalid directory entry
6	3	Invalid page-table entry
7		Protection fault
7	xxxxxx1x	Length
7	xxxxx1xx	Rights
9		Structural fault
9	1	Control
10		Type fault
10	1	Mismatch
12		Process fault
12	1	Time slice
13		Descriptor fault
13	1	Invalid STE
14		Event fault
14	1	Notice

The structural fault signifies a disastrous situation, almost always caused by a mistake in one of the architecture-defined data structures (e.g., a bad entry in the fault or interrupt table).

The type fault occurs if one attempts to change the process controls (as opposed to just reading them) when not in the supervisor state, or if the encoding of a segment-table entry is incorrect (e.g., when one references something other than a port entry in a SEND instruction).

The process fault occurs when the process hits the end of its time slice and the *tsr* flag in the process controls is clear. Like virtual-memory and protection faults, this fault can result in the suspension of the current instruction.

---

*It may seem strange to define the protection fault such that the fault handler can resume execution from the point of the fault. However, an operating system may wish to allow regions to "grow" on demand by dynamically expanding the size of the page table or directory as needed.

The descriptor fault occurs when there is a reference to an invalid segment-table entry. Typically this indicates an error in the operating system.

Finally, the event fault signifies an event signaled via the process-notice field in the PCB.

Along with the trace fault, the virtual-memory, protection, and descriptor faults are classified as *precise* faults because they are faults which the fault handler could be expected to want to correct and then resume execution; the others are imprecise faults. Imprecise faults can be made precise by setting flag *nif* in the arithmetic controls.

Overrides

There is one entry in the fault table (the first entry) that is not associated with a particular fault type; instead it is associated with a condition called override, which occurs whenever a second fault occurs while the chip is trying to store the fault record associated with a previous fault. The prime example of this situation is a virtual-memory fault that occurs while the processor is writing a fault record to the stack.

When this situation occurs, the first fault handler is not called; instead, the procedure specified in the override entry in the fault table is called. The fault record given to this procedure is the same as for the original fault, except that the information about the second fault is stored in a second set of fields in the fault record (OFlags, OType, OSubtype, and override fault data). In the saved process controls field in the fault record, the *res* (resume) and *rf* (refault) flags are set. This gives the software the information needed to correct the condition associated with the second fault, without losing track of the first fault.

The override fault handler can recreate the first fault, if it so chooses, by simply returning. If, upon a return from a fault handler, the processor sees that the *res* and *rf* flags are set, it generates another fault at this point, using the information in the fault record on the stack at the point of return as a description of the fault.

This mechanism has a secondary use - it allows software to simulate faults. If a procedure wishes to simulate a fault (which requires being in supervisor state), it can construct a fault record on the stack (just prior to its stack frame), set the *res* and *rf* flags in the process controls in the fault record, set the return status in register r0 to that of a fault handler, and return. In using this mechanism, one can also use additional software-defined fault types and additional software-defined subtypes of the hardware-defined fault types.

Tracing and Faults

In general, if a trace event is detected in an instruction in which another type of fault occurs, the latter fault will be taken and the trace event will disappear. For instance, if one has a hardware breakpoint register set to the address of an ADDI instruction and the ADDI instruction results in an overflow fault, the trace fault associated with the breakpoint will not occur.

This is not the case, however, with instructions that result in suspension and later resumption. For instance, a breakpoint on a LOAD instruction that results in a page fault will result in a trace fault eventually when the instruction is resumed and completed successfully.

System Errors

The override described in the previous section represents support for graceful degradation in the presence of multiple faults. The architecture defines two further levels of degradation. The first is the system-error interrupt. If a fault occurs while trying to invoke the override handler, the processor stores information about the latest fault in the PRCB and generates an interrupt with vector 248. The fault record that existed at the time of the latest fault is stored at offset 128 in the PRCB, and the type and subtype of the latest fault is stored at offset 72. Thus, in the case of a fault (1) that resulted in another fault (2) while invoking the fault handler that resulted in another fault (3) while invoking the override handler, information about all three faults is accessible in the PRCB.

The system error occurs in one other instance: when a fault occurs while there is no current process. This could occur because of a fault during dispatching, while resuming a process, or while executing an interrupt handler in the idle or stopped state.

The handler of a system-error interrupt should not execute a RETURN instruction, as the architecture does not guarantee what will happen in this circumstance.

One last possible situation is when yet another fault occurs while the processor is trying to invoke the system-error interrupt handler (e.g., the entry in the interrupt table is invalid or the interrupt stack is not mapped in accessible memory). The processor gives up at this point by entering the stopped state, but it also asserts its FAILURE pin, giving one a final indication that something has gone wrong.

The remaining information in this chapter is implementation dependent, that is, specific to the 80960MC but not necessarily to other implementations of the protected architecture.

INITIALIZATION

Initialization of the 80960MC is almost identical to that of the 80960KA as described in Chapter 2. Physical location 0 points to a segment table on a 64-byte boundary, location 4 points to a PRCB, and location 12 points to the first instruction.

During initialization, the eighth entry of the segment table is fetched. This entry is used to define the location and mapping of the segment table. It must be an entry describing a small (one page) or large (mapped by a page table) segment table. The segment table located via entry 8 need not be the same table containing this entry; in other words, it need not be the same table as pointed to by location 0.

The information fetched from the initial PRCB includes the processor controls. It is possible to specify in the processor controls of the initial PRCB that address translation is enabled, meaning that the processor will initialize with address translation enabled. In all cases, whether address translation is enabled or not, region 3 must be defined; its SS must point to some type of valid entry in the segment table. If one initializes with address translation enabled, it is important to realize that regions 0-2 have an unpredictable mapping at this point, since there is no process; therefore, the initialization code should not use virtual addresses below C0000000.

IAC MESSAGES

The 80960MC adds 13 IAC messages to those present in the 80960KA and KB. The new IAC messages are summarized below.

RESTART PROCESSOR	Field3 and field4 are physical addresses of a segment table and PRCB. The processor completely reestablishes its state by using these as the new segment table and PRCB. The next action is determined by the state field in the processor controls in the PRCB. That is, the processor enters the stopped, idle, or executing-a-process state. If it enters the executing-a-process state, the processor resumes execution of the process specified in the current process field in the PRCB.
WARMSTART PROCESSOR	Similar to RESTART PROCESSOR, except that the processor first stores the current processor controls in the current PRCB.
STOP PROCESSOR	The processor enters the stopped state. If it was in the executing-a-process state, it first suspends the current process.
STORE PROCESSOR	The processor controls are stored in the PRCB.

MODIFY PROCESSOR CONTROLS   Field3 is a value and field4 a mask. The processor controls register is assigned the value (field3 & field4) | (processor controls & ~field3). Changing the state field of the processor controls has an undefined effect.

FLUSH PROCESS   If the processor is in the executing-a-process state, the current process is suspended, any STE and page-table information for regions 0-2 is flushed from the TLB, and the processor resumes execution of this process.

CHECK PROCESS NOTICE   Field3 is a PCB SS or zero. If the processor is in the executing-a-process state, and if field3 is zero or the SS of the current process and the process is not in the interrupted state, the process-notice field of the PCB is fetched. If bits 16 and 31 of this word are set, they are cleared and an event fault occurs.

PREEMPT   If the processor is idle or executing a process not in the interrupted state, it examines the dispatch port. If it contains a process whose priority is greater than both that of the current process and of the nonpreempt-limit field of the processor controls, the current process is suspended and re-enqueued on the dispatch port, and the processor dispatches from the dispatch port.

FLUSH TLB   All page-table entries present in the TLB are flushed. All STEs present in the TLB, other than those for the regions and current PCB, are flushed.

FLUSH TLB PTE   Field3 is an offset into a segment and field4 is an SS. If there is a page-table entry in the TLB associated with this address, it is flushed.*

FLUSH LOCAL REGISTERS   Field3 is a physical address. If the processor is in the executing-a-process state, any onchip local register sets associated with a stack frame in the

---

*To flush a page corresponding to a particular virtual address, field4 should be the SS of the containing region and field3 is the virtual address with the upper two bits cleared.

page specified by field3 are flushed to memory, and all but the current register set are marked as purged (i.e., not contained onchip).

FLUSH TLB PHYSICAL PAGE   Field3 is a physical address. Any page-table entries in the TLB that point to the page containing the physical address are flushed. Any STEs that point to a page table or page-table directory in this page are also flushed (other than the region-3 STE). Also, the function of the FLUSH-LOCAL-REGISTERS IAC message is performed, and, if an STE for a region was flushed, the function of the FLUSH-PROCESS message is also performed.*

FLUSH TLB STE   Field3 is an SS. If there is an STE entry in the TLB corresponding to this SS (other than STEs to the current process and regions), it is flushed.

It is important to realize that these IAC functions have two purposes: to perform certain implementation-dependent functions in single-processor systems, and to perform certain control functions in multiple-processor systems. The only two functions that make little sense in single-processor systems are PREEMPT and CHECK PROCESS NOTICE.

There are three IAC functions that are associated with "restarting": RESTART, WARMSTART, and REINITIALIZE (the last was defined in Chapter 2). The first two differ from the third in that they can restart the processor in any state (stopped, idle, or executing a process). WARMSTART differs from RESTART in that the processor first writes the processor controls into the PRCB in memory; this is the only information in the PRCB that is not static.

There are two functions associated with "stopping" - STOP and FREEZE (FREEZE was defined in Chapter 2). STOP suspends the current process first; FREEZE does not. In a multiple-processor system, if one needed to do such things as move the segment table or PRCB, or change information in the PRCB, one could do so by telling each processor to stop and then to warmstart.

The MODIFY PROCESSOR CONTROLS IAC message allows one to change the processor controls register (in this or another processor), allowing one, for instance, to enable or disable address translation.

The FLUSH PROCESS IAC message causes the processor to write to memory all dynamic state information about the current process, and then to do the equivalent of a RESUMPRCS instruction for the process. For instance, if the operating system

---

*The reason for the flush-process is that the 80960MC does not dynamically fetch STEs for regions 0-2; it only does so when the processor resumes the process.

wanted to give the current process a new region 0 (or change the mapping of region 0 from paged to bipaged), it would change the region 0 SS in the PCB and then send the processor a FLUSH PROCESS IAC message. The parameter on this IAC message allows one to ask a processor to perform the function unilaterally, or only if it is executing a particular process.

The functions associated with TLB flushing are used when implementing paging. First, let us examine what must be done in a single-processor system to swap a page out of memory. We need to ensure that there is no stale data on the page (data that is not up-to-date), so we would use the FLUSH LOCAL REGISTERS IAC message.* Next, we need to make sure that the mapping information is removed from the TLB, so we would use the FLUSH TLB PTE IAC message. The final step before actually writing the page to disk, which is only necessary if its altered flag is set, is to mark the page-table entry as invalid, so that any future references to the page result in a virtual-memory fault.

In a multiple-processor system, the situation becomes more complicated in several ways. First, we must ensure that the information gets flushed from the TLBs of the other processors. Without the 80960's ability to send external IAC messages, we'd have to generate interrupts in all processors so that the operating system could locally flush each processor's TLB. Second, because the other processors are running concurrently and may be accessing the page as we're trying to swap it out, the order in which things are done is crucial. In the multiple-processor case, the first step is to mark the page-table entry as invalid, which will generate a fault in a program on any processor if it attempts to access the page and the page is not already in that processor's TLB. Next, if the page could contain part of a process's stack, the FLUSH LOCAL REGISTERS IAC is sent to all processors, along with the physical address of the page. Finally, the FLUSH TLB PTE IAC is sent to all processors to ensure that no one still has mapping information to the page.

In some cases a physical page can have aliases, meaning that multiple page-table entries point to the page. Here one has three choices for TLB flushing. The simplest is to use the FLUSH TLB function, but this is rather drastic in that it empties the entire TLB. The second is to send FLUSH TLB PTE messages for each virtual address by which the page is known. The third is to use the FLUSH TLB PHYSICAL PAGE function, which flushes any and all page-table entries and STEs pointing to the page.

There is no function associated with flushing page-table directory entries. The 80960MC does not place directory information in its TLB.

In Chapter 2, which discussed the mechanics of sending external IAC messages, we pointed out that an IAC message must be sent to a specific recipient. This seems undesirable, particularly in the situation of TLB flushing, where one would like to "broadcast" the request rather than send it individually to each processor.

---

*More typically, assuming the operating system always uses the same region for stacks, it would use this IAC message only if the page were part of this region.

Broadcasting would have two benefits: it would require the message to be sent only once, and it would not require the sender to know the identifiers of all other processors. It is therefore worth discussing why no broadcast capability exists.

At several points during the design of the processor architecture, BXU, and AP bus (which interconnects BXUs), we attempted to define a broadcast protocol, but found no satisfactory solution. One major problem is that IAC messages can be rejected, because the IAC buffer in the recipient holds a previous message, or because of priority. A broadcast protocol is useful in this circumstance only if one can identify who rejected and who did not, and no easy solution for this problem was seen. One solution might have been to eliminate the notion of rejection altogether, but this would have required a queue of IAC buffers in the BXU, with no clear way of establishing the queue depth and ensuring that the depth was not exceeded.*

Another problem is that often it is not sufficient to send an IAC message to a processor; one also needs a way of knowing when the IAC has been processed to avoid race conditions; for example, if we send a FLUSH PROCESS message to another processor, we need a way of knowing when it is done before we try to examine the PCB.

Given this last point, we should point out how one determines when another processor has finished processing a previous IAC message. When a processor reads an IAC message from the external buffer (a buffer in the BXU, assuming one is using BXUs), the buffer's status does not immediately change to empty; it is the processor's responsibility to write into the external control register associated with the buffer when the buffer is to be marked as empty.

For IAC messages that do not require any memory write operations (e.g., TLB flushes), the processor marks the buffer as empty right after it reads the message. For other messages (e.g., LOGOUT, FLUSH PROCESS, FLUSH LOCAL REGISTERS), the processor does not mark the buffer as empty until it finishes all write operations that result from performing the indicated function. One can then determine whether a previous IAC message has been processed by sending the processor a *second* IAC message. If the second message is not rejected, the first one has been processed. Unless one has a valid second message to send, the second message should have an undefined message type (e.g., 0), since the processor ignores any messages with an unknown type.

---

*A further complication is that IAC messages have an associated priority (part of the address field). If queues of IACs existed, one could argue that the BXU should then implement automatic priority ordering of the queue.

PREEMPTION IN MULTIPROCESSOR SYSTEMS

When one is using the automatic process-management model in a multiple-processor system, there is another optional facility for process management: *multiprocessor preemption*. Whenever preempting (typically, high-priority) processes are ready, multiprocessor preemption ensures that they are executed on some processor immediately, rather than waiting their turn in the dispatch port. This facility is triggered by flag *mpp* in the processor controls. Other pertinent information is flag *pr* in the process controls, flag *wep* and fields *interim-priority* and *nonpreempt-limit* in the processor controls, and the four-word field in the PRCB at offset 48.

A process, in addition to having a priority, can be marked as a preempting process (flag *pr*).* A preempting process is one that, in addition to having a high priority, needs immediate service whenever it is unblocked (becomes ready). Earlier in the chapter we saw that if a preempting process becomes unblocked, the processor switches to it if its priority is higher than that of the current process and puts the current process back on the dispatch port . If flag *mpp* is set, the following occurs instead.

We'll begin with a summary, because this mechanism is rather complicated. If a preempting process is unblocked (e.g., receives an awaited message), the processor unblocking it (e.g., executing the SEND instruction) compares the process's priority to that of the current process. If the unblocked process's priority is higher, the processor puts the current process back on the dispatch port and switches to the unblocked process, but if the process put back on the dispatch port is itself a preempting process, the processor notifies other processors that a preempting process on the dispatch port needs service.

If the unblocked process's priority is not higher, the processor puts it on the dispatch port and then tells other processors about its existence.

The processors notify other processors automatically, but this requires the cooperation of software in initializing the four-word field in the PRCB. When the notification is to be done, the processor fetches these four words and interprets them as a pair of memory addresses and words to be written to those addresses via the equivalent of a SYNMOV instruction. The intention is that these represent two one-word IAC messages (i.e., address and one-word message). Referring to them as messages A and B, the processor sends message A. If accepted, notification is complete. If not, message B is tried. If neither is accepted, they are tried a second time, but in this case the processor modifies the addresses by placing the priority of the process it is trying to notify other processors about in bits 4-8 of the address.

---

*Typically, one breaks the 32 priority levels into four classes. Level 0 is reserved to denote an idle processor. The next few low priorities are for nonpreempting priorities. Field *nonpreempt-limit* denotes the highest of these priorities. The next set of priorities is for preempting processes, and the highest set of priorities is for interrupts.

This mysterious algorithm is intended to do the following. Software would have initialized (statically) the four words so that the data words are PREEMPT IAC messages and the addresses are IAC addresses, each of which contains the identifier of another processor and a priority, the recommended priority value being 1. The algorithm then sends a priority-1 preempt message to another processor. If that processor is idle, it accepts it. Otherwise, the algorithm tries the second processor. If it is not idle, the algorithm tries the first processor again, sending it a PREEMPT message at the priority of the preempting process. If it is still unwilling, the algorithm tries the second processor. If this fails, it stops trying, with the expectation that, since the preempting process is on the dispatch port at the front of its priority queue, it will receive attention as soon as possible.

This manner of handling preempting processes obviously employs some heuristics and represents a design tradeoff. It does not guarantee that the n highest-priority processes will always run on the n processors, but in a probabilistic sense it tries to approximate this. The first heuristic could be called the "same processor" optimization. Independent of what the other processors are doing, if the preempting process has higher priority than that of the processor unblocking it, this processor runs it (with the motivation that trying to find a "better" home for it could be counterproductive, since doing so would delay the resumption of the preempting process). However, if by switching to the preempting process the processor preempts a lower priority preempting process, it is obligated to try to find a home for it first.

A major design tradeoff is the fact that a processor has only two "buddy" processors for preemption. This does not restrict systems to three processors; it just means that, for purposes of preemption, a processor notifies only two other processors. One reason for the tradeoff is the lack of a broadcast mechanism, as discussed earlier. Another is that trying a long list of processors could again be counterproductive in the case where the processor has decided to run a preempting process but first must try to unload the current, lower-priority preempting process.

Another heuristic is the two attempts that are made to preempt each buddy processor. If software initializes the PRCB information, as suggested above, so that the priority value in the addresses is 1, the processor first attempts to determine if either of its buddy processors is idle before attempting to preempt them at a higher priority.

The description of what happens when a processor receives a PREEMPT message pointed out that when the processor examines the dispatch port, it looks for a process whose priority is greater than both that of the current process and field nonpreempt-limit. The latter is done to ensure that the processor does not suspend the current process if the preempting process has already been dispatched by another processor (e.g., because another processor went to the dispatch port in between the time that the preempting process was enqueued there and the time that this processor examined the dispatch port).

As mentioned in Chapter 2, the processor writes its priority into the memory-mapped control register of the BXU whenever the priority changes, but only if flag *wep* in the processor controls is set. The 80960MC does this when a process is resumed, when it enters the idle state, when an interrupt handler is invoked, and when the execution of a MODPC instruction changes the priority. This priority determines whether the BXU accepts an IAC message or not. The BXU accepts a message (and signals the processor) if the buffer is empty and the message's priority is 31 or greater than that of the processor. Although every IAC message has an associated priority, all IAC messages, except INTERRUPT and PREEMPT, are typically sent at priority 31.

There are three other situations in which the processor writes a priority into the BXU; in these situations the value is that of the interim-priority field in the processor controls. The intention was that software would initialize this field to represent the highest priority of the preempting processes. The first situation is during execution of a SEND, SIGNAL, or SENDSERV instruction when a process is unblocked. The second situation is at the beginning of execution of a SCHEDPRCS instruction. The third is during the processing of a PREEMPT IAC. During these situations, the processor may be making a decision about whether to switch to a preempting process. If another processor is about to send PREEMPT messages on behalf of another preempting process, it is best that this other processor find a different processor to accept its PREEMPT message.

Finally, the description of the PREEMPT message states that a processor should ignore it if in the interrupt state (because, as explained earlier, a processor cannot suspend an interrupted process). This means that if a processor is trying to preempt its two buddy processors, the preemption is completely ignored if the two buddy processors happen to be executing interrupt handlers. To ensure that preempting processes do not go unnoticed longer than necessary in this event, the RETURN instruction also does something on behalf of multiple-processor preemption in the case of an interrupt return. If an interrupt return causes the processor to no longer be in the interrupted state, and if flag *cdp* or flag *mpp* is set, the dispatch port is checked, and *cdp* is cleared. If there is a process on the dispatch port with a priority greater than field nonpreempt-limit and that of the processor now, the current process (if there is one) is suspended and reenqueued on the dispatch port, and the processor dispatches.

SYNCHRONOUS LOAD/STORE INSTRUCTIONS

This set of implementation-dependent instructions (SYNMOV, SYNMOVL, SYNMOVQ, and SYNLD) described in Chapter 2 exist in the 80960MC, but with a few changes. Since the instructions are intended to be directed at specific physical addresses (e.g., for sending an external IAC message), they function a bit differently when address translation is enabled. For "normal" load/store instructions, address translation is logically done for each memory word accessed, so that if the operand spans a page boundary (i.e., the two parts of the operand are not necessarily contiguous in physical memory), both parts of the operand are accessed correctly. In the synchronous instructions, the A operand (an address) is treated differently; the address is translated to a physical address, and then the physical address is incremented if multiple words need to be accessed.

The processor forces alignment of the A operand address for two reasons. For SYNMOV and SYNLD, the two low-order address bits are cleared. For SYNMOVL and SYNMOVQ, the three and four (respectively) low-order address bits are cleared. One reason for the alignment is so that one is not lulled into believing that these instructions cross page boundaries correctly. The second, and more important, reason is that these instructions stop the pipeline and wait for a reply indicating whether the access was performed. The processor's local bus and the AP bus support multiword transfers, but only for transfers that do not cross 16-byte address boundaries. The processor is responsible for splitting memory accesses that cross 16-byte boundaries into two accesses. If this splitting were allowed for the synchronous instructions, the processor would have needed additional sequencing logic to wait for the two replies and then merge them into a single reply.

REFERENCES

1.   C. A. Alexander *et al*, "Translation Buffer Performance in a UNIX Environment," *Computer Architecture News*, Vol. 13, No. 5, 1985, pp. 2-14.

CHAPTER 5

Bus and External Signals

The three 80960 processors have a 32-bit multiplexed external bus with burst-transfer capability. In this chapter we explain the operation of the bus and the processors' other external interfaces, along with the motivation for the design decisions. For detailed timing specifications, one should consult the Intel datasheets; for information on the design of specific hardware configurations, see Intel's hardware reference manual.

BASIC BUS OPERATION

Unlike some other microprocessor buses, such as that of the 80386, the 80960 bus is multiplexed; addresses and data are transmitted on the same bus lines. The bus is multiplexed for four reasons:

1. Since the multiplexed bus can transfer from one to four successive words of data for every address, the principal benefit of a demultiplexed bus (the ability to transmit an address and a data word simultaneously) is of much less significance, because the percentage of bus overhead created by address cycles is reduced. In fact, had the possibility existed to devote 64 pins to the bus, we would have opted for a 64-bit multiplexed bus rather than a 32-bit demultiplexed bus.

2. Bonding pads occupy considerable space on the die; use of a demultiplexed bus would have therefore increased the die size (and thus the processor cost).

3. Worse, a large number of bonding pads inhibit the ability to shrink the die size to denser semiconductor processes. That is, the design becomes "pad limited" because pads must remain relatively constant in size for ease of bonding the pads to the leads in the package.

4. The auxiliary 82965 BXU (bus exchange unit) is provided for multiple-processor or fault-tolerant configurations. The BXU has two bus interfaces: the processor bus and the AP system bus, which is also a 32-bit multiplexed bus. Had the processor bus been demultiplexed, the BXU would have required a package with significantly more pins (about 180).

To discuss the operation of the bus, we must first define the states through which the bus passes. The bus is in a particular state for the duration of one cycle (e.g., a 50 ns period for a 20 MHz processor and 40 MHz clock signal). The states are

T_a Address cycle. An address is transmitted on the A/D lines.

T_d Data cycle. Data is driven for writes and sampled for reads. A T_d cycle during which the READY# input* is not asserted is called a wait state. For writes, this means that the processor should stay in the T_d state driving the same data in the next cycle. For reads, this means that the data is not yet available and that the processor should repeat the same T_d state in the next cycle.

T_i Idle cycle. The bus is idle.

T_x Dead or recovery cycle. To give the chips' bus drivers time to turn around (from sampling the external pins to driving them), a T_x cycle is required before a T_a cycle if the previous cycle was a T_d state for a read operation.

T_h Hold cycle. Use of the bus has been relinquished to another bus master.

T_{hr} Hold-request cycle. This state exists only if the processor has been configured as other than the "owner" of the bus, meaning that the processor needs to arbitrate for the bus. In this event, the processor is in the T_{hr} state while requesting the bus from another bus master.

When there are multiple masters on the same bus (e.g., two processors, a processor and a DMA device, a processor and a BXU), it is important to differentiate

---

*The # symbol is used to denote a signal that is active low; that is, READY# is considered to be asserted true if it is at low voltage.

between the states that the processor is in with regard to the bus and the states that the bus is in. The states defined above apply to the state machine in the processor's internal bus-control logic. The only states that apply to the bus itself are T_a, T_d, and T_i.

For read and write operations, the bus protocol supports the transmission of one, two, three, or four words (or parts thereof) of data. For instance, a four-word zero-wait-state read or write would result in the following sequence of states:

$$T_a\ T_d\ T_d\ T_d\ T_d$$

The successive words are associated with successively higher memory addresses. For instance, if the address transmitted during the T_a cycle is 003F0000, the four data words correspond to addresses 003F0000, 003F0004, 003F0008, and 003F000C.

Multiword burst accesses are restricted to those that do not cross 16-byte (four-word) address boundaries. In other words, a three- or four-word operation is permitted only if the low-order four address bits are zero. A two-word operation is permitted only if the low-order three address bits are zero. There are two reasons for this. First, if burst accesses were permitted on any word boundary, the bus control logic outside the processor would require a 30-bit counter, rather than just a two-bit counter.

Second, the BXU component permits the design of a system with one, two, or four memory buses, and the BXU interleaves accesses across the memory buses on 16-byte boundaries (to balance the traffic across the buses). For instance, in the configuration in Figure 5-1, AP bus 0 maps addresses xxx...xxx0xxxx, and AP bus 1 maps addresses xxx...xxx1xxxx. If the processor were permitted to perform burst accesses across 16-byte boundaries, the two BXUs would have to split the access, each BXU handling part of it, and then coordinate the timing of the replies to the processor, a situation of considerable complexity.

In the protected architecture level of the 80960MC, where it is possible for a program to execute a multiword load or store instruction that results in a 16-byte boundary crossing, the processor handles the crossing transparently to software; it generates *two* accesses, neither of which cross the boundary.

Why is the maximum burst access four words as opposed to, say, sixteen words? The answer has several parts. First, as will be discussed in Chapter 8, the optimal line size for the instruction cache was found to be four words, meaning that if the bus supported larger burst accesses, they would be used very infrequently (primarily for saves/restores of local register sets and for process switches). Second, larger transfer sizes have diminishing returns. For instance, assuming zero wait states, if the transfer size were limited to one word, the latency to read four successive words would be 11 cycles, as shown below:

$$T_a\ T_d\ T_x\ T_a\ T_d\ T_x\ T_a\ T_d\ T_x\ T_a\ T_d$$

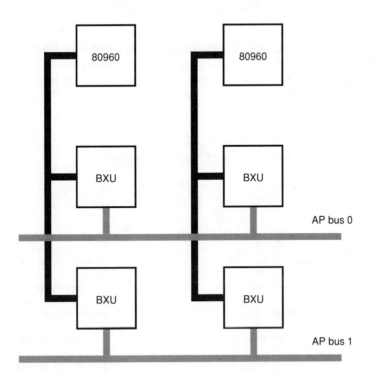

Figure 5-1: A configuration with two memory (AP) buses

With a four-word burst size, the latency is five cycles, or a reduction of 55%:

$T_a\ T_d\ T_d\ T_d\ T_d$

However, the savings are proportionately less for larger accesses. The latency to read 16 successive words is 23 cycles using four-word burst accesses, and 17 cycles using a hypothetical 16-word burst access, a reduction of 26%. Given that this lower saving would occur much less frequently, the overall benefit is insignificant.

The third reason for a maximum burst access of four words is that the processor contains three write buffers to avoid delaying write operations when the bus is busy. Each buffer is five words in size (an address and up to four words of data). If a larger burst-transfer size were supported, the write buffers would have consumed an excessive amount of silicon.

PIN DEFINITIONS

Figure 5-2 shows the external signal pins (i.e., all except V_{CC} and ground). Five of the pins (HOLD/HLDAR, HLDA/HOLDR, IAC#/INT0#, INT2/INTR, and INT3#/INTA#) have dual user-selectable definitions. These will be discussed as if they were individual pins.

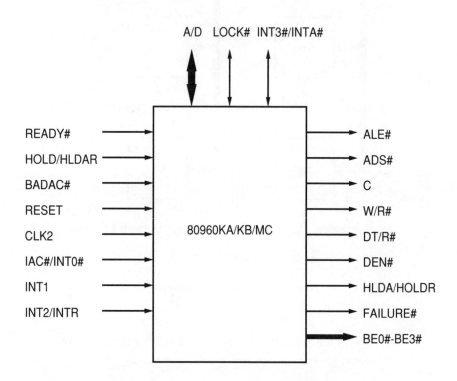

Figure 5-2:   80960KA/KB/MC External Pins

All pins that are outputs (or inputs/outputs) are driven by one of three types of drivers:

Buffer   The output is driven high or low at all times.

Open-drain   Only active-low outputs fall into this class. When the signal is
                     not asserted, it floats.   An external pull-up resistor is used
                     (typically 100 ohms to minimize rise time). A terminating pull-
                     down resistor (typically 390 ohms) should also be used to mini-
                     mize noise.

Tristate   When the output is defined to be valid, it is driven to a high or
                     low state. During states when the output is not defined, it floats to
                     a high-impedance state.

   Unless otherwise noted, active-low outputs are open drain and all others are
buffered.

Pins

A/D   32-bit address/data bus. During a T_a cycle, lines 2-31 contain the
                     address of a word, and lines 0-1 specify the number of words (1-
                     4) to be transferred. During a T_d write cycle, a data word is
                     driven. During a T_d read cycle, a data word is sampled at the end
                     of the cycle if the READY# input is asserted. The A/D lines are
                     tristate.

ALE#   Address-latch enable. ALE# is asserted at the beginning of a T_a
                     cycle, is deasserted (high) a half-cycle later, is held deasserted
                     during T_d cycles, and floats (tristate) at all other times. ALE# is
                     normally used to control an external address latch.

ADS#   Address strobe. ADS# is asserted during a T_a cycle and during
                     every T_d cycle for a subsequent word if the READY# input was
                     asserted during the previous T_d cycle. A T_a cycle can be identi-
                     fied by ADS# & ~DEN#.

C   Cacheable. During a T_a cycle C denotes (to an external cache)
                     whether the data being read or written is considered cacheable.
                     For example, when address translation is enabled in the
                     80960MC, the pin denotes the value of the C flag in the page-
                     table entry. C floats (tristate) when the processor is not using the
                     bus.

W/R#   Write/read. During T_a and T_d cycles this pin specifies whether
                     the operation is a write (high) or read (low). W/R# is an open-
                     drain output.

DT/R#   Data transmit/receive. This open-drain output is asserted (low) at
                the middle of a T_a cycle for a read operation and remains asserted
                until the middle of the cycle following the last T_d cycle. It is
                typically used to control the direction of external data-bus
                transceivers.

DEN#   Data enable. DEN# is asserted during T_d cycles.

BE0#-BE3#   Byte enables. These signals specify the bytes on the A/D bus that
                are part of the transfer. For instance, for a write operation corres-
                ponding to a STIB (store integer byte) instruction, only one of the
                byte-enable pins would be asserted.

                The BE signals are provided in advance of the data transfer. For
                the first (or only) word, they are asserted during the T_a cycle. For
                subsequent words, they are asserted during T_d cycles, where they
                specify the bytes of the next word.

                For a system design that is entirely 32 bits, the BE signals typi-
                cally are used only for write operations (where they specify the
                bytes of the data word that should be written). For a system
                design that contains a smaller (8- or 16-bit) memory-mapped I/O
                bus, the BE signals may be used for reads as well to denote the
                number of bytes that must be sequenced into buffers driving the
                A/D bus.

LOCK#   Bus lock. LOCK# is asserted in the T_a cycle of an RMW
                (read/modify/write) read operation, and is held asserted until the
                T_a of the corresponding RMW write operation. For instance,
                LOCK# is asserted by the processor while executing ATADD and
                ATMOD instructions, when obtaining a pending interrupt from the
                interrupt table, when setting the lock in the dispatch port, and
                when setting altered and accessed flags in page-table entries. If
                the system design has other bus masters that also perform RMW
                operations (e.g., another processor), LOCK# signifies that no
                other bus master should perform an RMW operation.

                LOCK# is also an input to the processor. When the processor is
                about to start an RMW read operation, it samples LOCK#. The
                processor does not enter the T_a state until it sees that LOCK# is
                not asserted.

READY#   READY# is an input sampled by the processor at the end of T_d cycles. It signifies that the processor can proceed to the next state (i.e., for writes, that the next word should be driven or that the operation is complete; for reads, that the data on the A/D bus can be sampled).

BADAC#   Bad access. BADAC# is sampled in the cycle following the last cycle of an operation (that is, following the one in which READY# is asserted signifying the end of the transfer). For normal operations, an asserted BADAC# results in the bad-access machine fault. For the synchronous load/store operations, BADAC# causes the condition code to be set. BADAC# is typically used to signal memory parity errors or unrecoverable ECC errors.

FAILURE#   The continuous assertion of this output signifies that the processor has failed its internal self-test, has failed the checksum of the first eight words of memory, or has entered the stopped state because of a series of fault conditions. Typically, FAILURE# is used to drive an LED. By design of a simple state machine to watch how FAILURE# is toggled after reset, one can distinguish among these three conditions.

INT0#   Interrupt request. If this pin is deasserted for one or more cycles and then asserted for one or more cycles, the processor interprets it as an interrupt signal. The vector used is the low-order byte of the interrupt-control register.

                Unlike all other pins, this and the other interrupt pins can be treated as synchronous or asynchronous. A synchronous input is one that obeys setup- and hold-time specifications with respect to the clock edge on which the input is sampled; an asynchronous input is one that does not. If used asynchronously, the pin must be deasserted for two or more cycles and then asserted for two or more cycles.

INT1   Interrupt request. This input is similar to INT0#, except that it is active high and corresponds to a different vector in the interrupt-control register. The interrupt inputs were provided in a mixture of active-high and active-low because there is no convention for the polarity of interrupt signals from peripheral devices (and thus one is usually faced with both polarities).

INT2   Interrupt request. Similar to INT1.

INT3#   Interrupt request. Similar to INT0#.

INTR   Interrupt request. Similar to the above, except that it signifies that
                  the processor should perform an interrupt-acknowledge bus oper-
                  ation to obtain the interrupt vector from an external device.

INTA#   Interrupt acknowledge. This is an output used to signal an inter-
                  rupt-acknowledge bus operation.

IAC#   IAC message request. This input signals the processor to read
                  and process an external IAC message.

HOLD   Hold request. This input requests that the processor relinquish
                  control of the bus to allow another bus master to perform an oper-
                  ation. The processor enters the T_h state immediately, or imme-
                  diately after finishing the current operation, and asserts the HLDA
                  output. The processor stays in the T_h state until HOLD is not
                  asserted. When HOLD is asserted, the processor enters the T_h
                  state in the next cycle if it is not using the bus, or immediately
                  after the last cycle of the current operation if it is using the bus.
                  When HOLD changes from asserted to not asserted, the next
                  cycle is a T_i cycle (if there is no operation pending) or a T_a cycle.

                  In many system designs, even in those with other bus masters
                  (such as DMA channels), it is possible to perform memory and
                  bus arbitration externally without using any of the processor's
                  arbitration signals. If the processor's A/D bus and control signals
                  are isolated from the memory with address latches, transceivers,
                  and control logic, one can use external logic to do the arbitration
                  and use the READY# input to delay the processor.

HLDA   Hold acknowledge. This output is asserted whenever the pro-
                  cessor is in the T_h state. It signifies that control of the bus has
                  been relinquished to someone else.

HOLDR   Hold request output. The system designer can configure the pro-
                  cessor so that it is not the defacto "owner" of the bus, which
                  means that it must request control of the bus prior to every opera-
                  tion. In this type of configuration, the processor asserts HOLDR
                  when it needs the bus, and enters the T_{hr} until the HLDAR input
                  is asserted, at which time the processor enters the T_a state (and

keeps asserting HOLDR). In the cycle after the last cycle of the operation, the processor deasserts HOLDR.

In general, it is unwise to configure the processor to use HOLDR and HLDAR, since this type of system design adds considerable overhead to every bus operation. These signals exist to allow one to place two processors directly on the same external bus, in which case HOLDR of processor B is tied to HOLD of processor A, and HLDA of processor A is tied to HLDAR of processor B. However, in addition to the arbitration delays that are experienced by processor B, both processors operate more slowly because of bus interference.

HLDAR
Hold request acknowledge. When the processor is not configured as the bus owner, this input signifies that the processor can exit the T_{hr} state and perform a bus operation.

RESET
This input tells the processor to reset. Because the use of this pin requires special care with respect to its timing relative to the clock signal, and since configuration information is sampled from other pins during reset, the reset operation is discussed separately in the next section.

CLK2
CLK2 is a 2x clock signal (i.e., if the processor is operating at 20 MHz, CLK2 is a 40 MHz signal). Bus cycles begin on every other rising edge of CLK2. Which specific rising edge is used is determined by the timing of the RESET signal, as described in the next section.

When designing a system, it is important to note that, with two exceptions, all of the output pins are clocked signals and are not guaranteed to be valid across cycle boundaries. For instance, during a burst read operation, the W/R# is driven low during all T_a and T_d cycles, but it may experience a short state change on each cycle boundary. The two exceptions are ALE# and DEN#, which are latched within the processor and held steady across cycle boundaries.

PROCESSOR RESET

In addition to resetting the processor, the RESET pin determines the clock edge on which bus cycles begin, causes the processor to read some configuration information from other pins, and initiates a self-test and toggling of the FAILURE# pin.

A reset is initiated by asserting RESET and holding it asserted for at least 32 CLK2 cycles.* When RESET is deasserted, the next rising edge of CLK2 (and every second rising edge thereafter) is the edge that begins a bus cycle.

When RESET is deasserted, the processor samples configuration information from two pins: IAC#/INT0# and BADAC#. In almost all cases, one should pull these pins high at reset to specify that the processor is the bus owner and an initialization processor. If IAC#/INT0# is pulled low, the processor is not the bus owner; in this situation pin HOLD/HLDAR will be interpreted as HLDAR and pin HLDA/HOLDR is interpreted as HOLDR. Whether or not the processor is the bus owner also establishes the processor's number with regard to the addresses it uses to read external IAC messages; a processor configured as the bus owner is processor "0".

If the BADAC# input is pulled low at reset, the processor will stop after performing its self-test, rather than performing a software initialization. The motivation for this provision is that in multiple-processor systems, it may not be desirable for all processors to perform the software initialization.

As described in Chapter 2, the processor performs two types of tests prior to software initialization: an internal self-test and a basic confidence test of the bus and memory. After the RESET pin is deasserted, but before the self-test begins, the FAILURE# output pin is asserted. If the self-test passes, FAILURE# is deasserted. If the checksum of the first eight memory words fails, FAILURE# is asserted again, and remains asserted. If the checksum is correct but the condition of having too many successive software faults occurs at any later time, FAILURE# is asserted and remains asserted. Thus, if desired, one could include a simple state machine (e.g., a PAL) that drives three LEDS instead of one to indicate which of the three "FAILURE#-asserting" situations occurred.

ELECTRICAL CHARACTERISTICS

The specification of signal timing (e.g., output delays, setup and hold times) is more an art than a science, and the semiconductor manufacturer must find a happy compromise between the benefits of tight specifications and those of loose specifications. Loose specifications (e.g., longer output delays and setup and hold times) benefit the manufacturer in two ways. First, they mean that considerably less time need be spent on circuit design, which in turn means less risk, less design cost, and faster time to market. Second, given the variations that exist from run to run in semiconductor fabrication plants, looser specifications result in higher yields (less cost).

---

*The state of the processor is not guaranteed if RESET is asserted for fewer than 32 cycles.

On the other hand, tighter specifications primarily benefit the user (system designer) by giving him or her more nanoseconds to play with, which leads to the potential benefits of a cheaper system design, faster design time, and higher performance. Tighter specifications also benefit the processor manufacturer indirectly by making the processor more competitively attractive.

With this bit of economics out of the way, we can now explain the approach used in the 80960. Tighter specifications were used (i.e., the system designer was favored). Any resultant loss in manufacturing yield was minimized by spending a considerable effort during the chip development on circuit design. Since the 80960 and 80386 were developed at roughly the same time by the same company, and since the 80386 took the path of looser specifications, it is worthwhile to examine the resultant differences in signal timing.

Since, with few exceptions, the input and output signals of both processors are synchronous to the input clock, the specification of the clock input signal has a significant bearing on the specifications of all other signals. The 80386 requires what is informally named a "CMOS" clock input, that is, a clock signal with a large voltage swing. The swing is defined in terms of V_{CC}; for a 5.0V V_{CC}, the clock low and high logic-level values are 0.8V and 4.2V, and the rise and fall times measured between these points must be 8 ns or less (for 16 MHz operation). For the 80960 processors, the clock low and high logic-level values (for 5.0V V_{CC}) are 1.0V and 2.75V, and the clock-signal parameters are defined with respect to 1.175V and 2.575V. At 16 MHz (32 MHz clock period), the clock signal must be high for at least 11 ns and low for at least 11 ns, leaving 9.25 ns (31.25 - 11 - 11) for the rise and fall between 1.175V and 2.375V. The result is that the 80960 processors are specified for a "sloppier" clock signal.

The reason for the difference is that the 80960's, particularly the 80960MC, were designed to support large multiprocessing configurations, where the designer must provide all processors and BXUs with a clock signal (e.g., at 32, 40, or 50 MHz) with minimum clock skew. The TTL clock specification makes this easier to achieve, although the added delay in the chips' clock buffers makes the set-up and hold times of input signals different than in the 386.

The specifications that are of most interest to the system designer are the delay in the T_a cycle until the address and control signals are valid, and the time prior to the end of the T_d cycle (the setup time) that the returned data must be valid. The duration between these two values forms the access time, that is, the time available to perform a memory access (unless one wants to add wait states).

The 16 MHz commercial temperature-range 80960KA and 80960KB are specified with a V_{CC} of 5.0V ±10% and capacitive load of no more than 100 pF on the A/D bus and 75 pF on the other control signals. The maximum output delay of all signals (e.g., the address) is 35 ns, and the minimum setup time of the data input is 3 ns. For zero wait-state operation, the access time available is 125 - (35 + 3), or 87 ns. Since typical system designs demultiplex the bus externally using address latches and data transceivers, another 15-20 ns disappear in propagation time through these devices. The end result is that it is reasonably easy to design a zero-wait-state system with an external cache.

As mentioned earlier, the 80960's went the route of tight specifications. This can be contrasted with the 80386, which has a maximum output delay of 38 ns and a minimum setup time of 10 ns, giving the designer 77 ns of access time.

At 20 Mhz, the 80960KA and 80960KB are specified as requiring a V_{CC} of 5.0V ±5% and 25 pF less capacitive load. The maximum output delay is 30 ns, giving the designer 67 ns of access time (100 - 30 - 3). Even after subtracting the propagation time through external latches and transceivers, a zero-wait-state cache can be achieved by use of 35 or 45 ns static RAMs.

As will be discussed in Chapter 8, the 80960K-series processors have several attributes that make them much less sensitive (in terms of performance) to wait states than is the case with other processors. Cacheless DRAM designs are therefore often a better cost/performance compromise. Table 5-1 shows the bus timing that one could expect to achieve with different types and configurations of 120-ns DRAMs. The timing shown is for a four-word read, where "-" represents a wait state. The "normal" configuration consists of one or more noninterleaved DRAM banks. The "interleaved" configuration assumes two interleaved DRAM banks. The "static column" configuration assumes noninterleaved DRAM banks using a type of enhanced timing known as static column decode or fast page mode found in many DRAMs, where successive words can be accessed at a rate of 60 ns or better per word.

Table 5-1: Bus timing for 120 ns DRAMs

Configuration	Freq	Bus cycles
Normal	16	A - D - - D - - D - - D
Interleaved	16	A - D D - D D
Static column	16	A - D - D - D - D
Interleaved	20	A - - D D - D D
Static column	20	A - - D - D - D - D
Static column, interleaved	20	A - - D D D D

INTERRUPTS

The processor contains four pins associated with interrupts. Depending on the way the interrupt control register is initialized, the pins can be used for

- four interrupt signals

- three interrupt signals and one external IAC message signal

- two interrupt signals and two handshaking signals to an external interrupt controller

- one interrupt signal, one external IAC message signal, and two hand-shaking signals to an external interrupt controller

When the pins are configured to handshake with an external interrupt controller, one pin is defined as INTR and the other as INTA#. The protocol and timing used assume that the external interrupt controller is an Intel 8259A. When an interrupt is signaled on INTR, the processor performs two read operations. The address used is FFFFFFFC, and output INTA# is asserted for the duration of each of the read operations. The processor inserts five idle states before the second read to adhere to the timing conventions of the 8259A. The low-order byte transmitted during the second read is assumed to be an 8-bit interrupt vector number.

The rationale for the two reads is the way the 8259 was designed (many years ago) to support both the 8085 and 8086. The rationale for the INTA# signal is that it is used to trigger the 8259, and is typically used in the system design to distinguish interrupt-vector read operations from normal read operations. The rationale for the address FFFFFFFC is in BXU-based systems; we could not afford to place an INTA# input pin on the BXU, and thus the BXU treats FFFFFFFC as a dummy address and does not attempt to respond to it.

Use of the 8259 presents three problems with respect to the priority-interrupt mechanism of the 80960 architecture. The first is that the 8259 returns interrupt vectors of the form ccccciii, where ccccc is a constant programmed into an 8259 register and iii is the encoding of a specific interrupt line. Since the 80960 architecture interprets the upper five bits of the vector as the priority, all the interrupts from the 8259 have the same priority level.

Although this may be exactly what is desired in some system designs, it is more likely that one wishes the interrupts to have different priorities. This can be accomplished by connecting the data bus of the 8259 so that the vector is logically rotated when delivered to the processor. For instance, the 8259's data bus might be connected in such a way that the vector appears as cciiiccc to the processor. This enables one to have the eight interrupts be of priority 0-7, 8-15, 16-23, or 24-31, depending on the value of ccccc that is stored in the 8259 at initialization.

The second problem is that the 8259 contains priority-resolution logic which is unfortunately opposite of the priority encoding in the 80960 architecture. For instance, 8259 interrupt input pin IR7 corresponds to iii=111, and pin IR0 corresponds to iii=000. If multiple interrupts are present in the 8259, the one sent in the vector is the lower numbered one, which, if one rotates the bus as suggested above, causes the lower priority interrupt to be signaled first. This can be corrected by using an 8-bit inverting buffer between the 8259 and the data bus.

If the eight-bit bus to the 8259 is both rotated and inverted, one must remember that this applies not only to the interrupt vector, but to any 8259 control-register values that are read or written.

The third problem is that the 8259 is not a clocked device and has no reset signal; therefore it may generate one or more spurious interrupts by chance when power is applied. Since the processor masks interrupts only by priority and has no means to mask all interrupts completely, the suggested solution is to put an AND gate between the 8259's INTR output and the processor's INTR input, and tie the other AND input to the output of an external control register whose output is guaranteed to be 0 at reset. The initialization software can then enable the 8259 by changing the control register after it initializes the 8259's registers.

CHAPTER 6

The Implementation

This chapter describes the first implementation of the 80960 architecture. Since the 80960KA, 80960KB, and 80960MC were derived from a single design, the discussion of the microarchitecture will primarily focus on the 80960MC implementation.

The 80960MC (as well as the 80960KA, and 80960KB) is implemented using Intel's CHMOS III, two-layer metal, 1.5-micron gate length technology. The chip measures 388 by 396 mils and contains more than 350,000 transistors. The internal structure, data and control flow, instruction pipelining and other performance techniques, and test features are described here. The rationale for many of the decisions and tradeoffs that were made in the design process are also discussed.

This chapter does not provide a comprehensive discussion of the implementation. We have tried to point out the important structures and their interactions with other parts of the chip. The main motivation is to provide enough detail so that the reader can appreciate the design decisions that were made in the implementation and, when reading the other chapters, associate implementation issues with the corresponding architecture features and performance.

OBJECTIVES

The initial design objectives of the first implementation were set in mid-1982. It was recognized that high performance was the initial entry requirement to establish a new architecture in an environment in which existing microprocessor families were already evolving to 32 bits. Examples of these included the Motorola 68000 and Intel iAPX-86 families. These families were already established in several markets, and performance was the clear metric in which these processors competed. In addition, the 32-bit members of these microprocessor families were beginning to approach the performance of existing super-minicomputers such as Digital's VAX. In the performance arena, the VAX 11/780 was the machine against which all others were measured. The global performance goal for the first implementation was to provide three times the power of a VAX 11/780 for single processor performance across all benchmarks in realistic system environments,[*] not just benchmark systems. Other, more specific goals included running the Whetstone floating point benchmark at greater than 3 MWhetstones/second, providing a sustained bus bandwidth of greater than 40 Mbytes/second, and providing support for multiprocessing.

The second goal was to provide a feature set that would allow the chip to solve system problems that were not being addressed with the existing product families. This would be accomplished by providing a flexible system architecture that would allow the construction of large multiprocessing systems with multiple buses and offer a wide range of system configuration and fault tolerant capabilities. Systems could be constructed with an initial feature and performance level and later be upgraded in performance and capability.

Another goal was the full integration of the memory management and floating point functions into the processor chip. Floating point support has traditionally been supported through the use of coprocessor chips. This usually presents a performance bottleneck because of the processor to coprocessor communication required to pass operands and return results. The communication overhead can easily derate the processor performance by more than a factor of two for floating-point-intensive applications. It also is less cost effective when the overhead in duplicated logic is considered: a coprocessor would require a bus interface and microcode sequencing logic similar to that already designed into the processor chip. Some product families (Motorola 68000 and National 32000) also require additional chips for the memory management functions which presents additional delays to memory, in the form of extra wait states, typically one extra wait state for each memory access. The Intel 80286 and 80386 devices are examples of chips that fully integrate this function into the CPU.

Finally, considerable emphasis was placed on defining an architecture that would be subsettable. As described in the previous chapters, there are three levels of the 80960 architecture. While the first implementation supports all three levels, it was desired to be able to proliferate many future implementations tuned to the needs of specific applications while providing a common architecture that would allow a high degree of software compatibility.

---

[*]This goal was later determined to be too conservative, and the resulting performance goals were increased by a factor of two over the development period.

Several constraints or boundary conditions were placed on the development of the chip. These included limiting the die size to 400 mils per side, maintaining a power dissipation less than 3.5 watts, a microcycle target of initially 16 MHz with later versions at 20 and 25 MHz, and a 3.5 year development cycle from concept to working samples.

The die size limitation was driven primarily by equipment capabilities for the process at the time of initial manufacturing (late 1985). The lens of the wafer stepper used in the production process flow had a field size limitation of 400 mils. Product cost was a secondary concern. While this is typically the most important issue, the 80960KA/KB/MC was a new product in a new product family that over the years would evolve and proliferate. Volume requirements in the early years of its life would be small (thousands rather than hundreds of thousands). The traditional learning curve and product evolutionary cost reduction practices would suffice to meet the later high volume production requirements.

The cycle time was both a goal and a constraint. As a goal, it determined the raw performance of the basic operations (e.g., register to register operation, internal bus transfer times, and external bus timing). As a constraint, it defined the boundaries for all circuit design activity, such as the clock generation and distribution, ROM, cache and register file access times, and the total time per phase for logic evaluation.

80960MC OVERVIEW

The chip is comprised of seven major units as shown in Figure 6-1.

The *instruction decoder* decodes and controls instruction execution. It performs effective address computation and operand fetching, executes branch instructions (i.e., instruction pointer manipulation), and either emits execution microinstructions (for simple instructions) or starts microprogram flows (for complex instructions). The microinstructions are bit patterns that typically look very similar to instructions, but in many cases contain additional information such as the address of the next microinstruction to be issued.

The *instruction fetch unit* fetches, prefetches, and caches instructions from memory to keep the instruction decoder supplied with a stream of instructions. The instruction fetch unit also maintains six instruction pointers that track instructions through the pipeline.

The *integer execution unit* contains the four sets of local registers, the global registers, a set of scratch registers used by microcode, the arithmetic logic unit (ALU), the barrel shifter, and the logic needed to execute the 32-bit arithmetic and logical instructions.

The *floating point unit* contains the floating point registers and the necessary logic to execute all the floating point instructions and integer multiply and divide instructions.

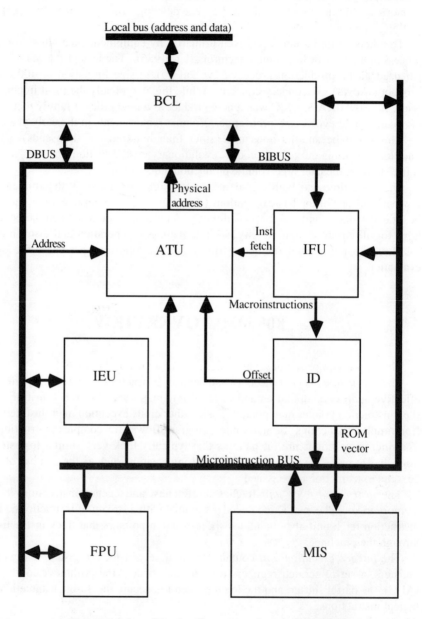

Figure 6-1:   Block diagram of the 80960MC

The *address translation unit* performs address translation and memory protection using an associative table of storage descriptors implemented as a translation lookaside buffer (TLB).

The *microinstruction sequencer* contains the microcode ROM, fetches the next microinstruction, controls microprogram branching, maintains a scoreboard on the register file, and handles exception and interrupt conditions.

The *bus control logic* interfaces to the external bus, provides a pipelined buffer for several memory requests, manages the bus protocol, and recognizes external events (e.g., interrupts, initialization).

Global interunit communication is accommodated by two 32-bit data buses (the DBUS and the BIBUS), a 29-bit microinstruction bus, and a microinstruction valid signal.

Instruction Pipelining

In the most common instruction sequences, namely register operations, the processor has a five-stage pipeline, meaning that five instructions can be processed concurrently. As an example, the following can occur in the same cycle:

- The integer execution unit is storing the result of instruction n.

- The integer execution unit is fetching two operands (one from the bypass path, or a literal value) and performing an ALU operation on behalf of instruction $n+1$.

- The microinstruction is issued on the microinstruction bus for instruction $n+2$.

- The instruction decoder is decoding instruction $n+3$.

- The instruction fetch unit is accessing instruction $n+4$ from the instruction cache. And, possibly, it is doing a prefetch of another cache line.

Memory operations are also pipelined. If a program is executing an instruction to load an operand from memory, the processor starts the memory read by assembling an address and issuing it to the external bus along with the appropriate control information. The destination registers are scoreboarded (i.e., marked invalid). The processor then proceeds immediately to the next instruction before the data is returned from memory. When an instruction is executed, it proceeds providing that its source and destination registers are valid; otherwise the instruction is retried. When data is returned from memory, the destination registers are marked valid. In this way, the processor can execute instructions in parallel with the memory operation until it hits one that references an invalid register. If an instruction occurs that requires the data from the prior load instruction, it *spins* (loops) until the data is available. The spin is accomplished by deactivation of the microinstruction valid signal by the scoreboard logic. This, in effect, causes the microinstruction to be reissued and inhibits any impending state change. Store operations are handled in a similar way, but without scoreboarding.

The motivation for this scoreboarding can be seen in the following example:

(A)		(B)	
ld	(r7), r1	ld	(r7), r1
addo	r1, 1, r1	addo	r2, r3, r4
addo	r2, r3, r4	addo	r1, 1, r1

In code sequence A, a value is loaded into register r1 from the location specified in register r7. Register r1 is then incremented, and registers r2 and r3 are added and stored in register r4. In this case, the second instruction hits the scoreboard and causes instruction execution to wait for the data from the load instruction to be returned from memory.

In sequence B, the second two instructions are reordered to take advantage of the scoreboard. Since the second instruction in sequence B does not depend on the data in register r1, it can proceed in parallel with the memory operation. This code reordering is an easy optimization for compilers to do.

The bus control logic also contains a FIFO (first-in/first-out) buffer (Figure 6-2) in its output data path for buffering memory operations so that instruction execution can proceed independently of the operation of the external bus. For example, if a load is issued followed by a store (one that doesn't hit the scoreboard) the following occurs:

- The address of the load instruction is issued to the address translation unit for translation and then to the bus control logic and on to the external bus (assuming the bus is idle).

- The destination register for the load is scoreboarded. Control then proceeds to the store instruction.

- The address of the store is issued to the address translation unit for translation and then to the bus control logic. This time the bus is not idle, so the address and subsequent data are held in the FIFO waiting for an available bus cycle. In the meantime, the processor is free to continue execution.

The FIFO can buffer up to three memory operations. When the buffer is full, the bus control logic deactivates the microinstruction valid signal and causes the processor to spin in a manner similar to that described above until a free entry is available.

The bus control logic contains a similar FIFO in its input data path. This allows buffering data from memory until it can acquire the microinstruction bus to transfer data to the destination registers.

The net effect of the scoreboard and the bus control FIFOs is that memory operations do not inhibit instruction execution throughput to the same degree as in other microprocessors.

Asynchronous Events

Asynchronous events are those that implicitly alter the sequence of microinstructions. Five types of events are supported:

- *Injected microinstructions*. The microinstruction bus is a shared resource onto which different units can place microinstructions. The normal uses of the bus are (1) by the instruction decoder, which places microinstructions on the bus to execute simple instructions and (2) by the microinstruction sequencer, which places microinstructions on the bus from the ROM. The bus control logic and address translation unit can interrupt these microinstructions by injecting their own microinstruction on the bus (primarily to move data to or from the integer execution unit's register file). These injections are transparent to the normal microinstruction flow (except for the delay and an occasional perturbance in the integer execution unit's bypass path, leading to loss of a cycle).

- *Traps*. A trap is an event caused by the detection of an error, most of which become faults. A trap causes an implicit microcode call to a predefined ROM location.

- *Microassists*. Microassists are events that are caused when a condition detected by a unit (e.g., floating point unit, address translation unit) requires a function be be performed that is not implemented in hardware logic (i.e., microcode is expected to handle the condition). Microassists are handled in the same way as traps.

- *Interrupts*. Interrupts are events resulting from the receipt of external interrupts and IAC messages. Unlike traps and microassists, interrupts may be enabled and disabled.

- *Instruction boundary events*. These are events that can occur when an "end of instruction" state is reached or when the processor is idle. For example, these events occur at the end of every trace boundary when the tracing mechanism is being used.

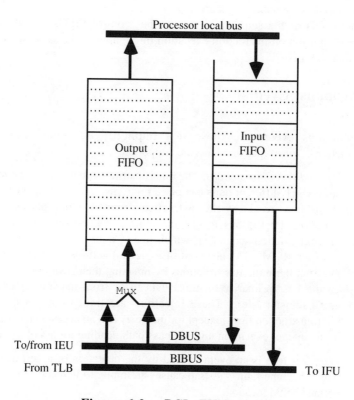

Figure 6-2:   BCL FIFO structure

Testing Features

The chip contains a set of mechanisms that enable it to be easily tested. The key mechanism is the ability to execute special test instructions from external memory.

This mechanism allows the processor to fetch test code from memory and place it in the instruction cache, just as it would if it were executing instructions, but the processor subsequently interprets the bits as special test instructions rather than as instructions. This capability allows one to generate test programs to reach the internal logic of the chip in ways not usually accessible by use of the normal instruction set. Special test instructions have been defined to allow access to control registers and state information. Also, the chip is logically partitioned so that, as much as possible, each unit can be tested individually. Of course, the unit tests assume that the major buses, clocks, and register file are working, but these are all checked by the self test microprogram described below.

Other functions provided for test purposes include

- *Self test microprogram.* One hundred words of internal microcode were allocated for an internal self-test microprogram. When the processor is reset, the self-test microprogram is executed and the FAILURE pin is set to the asserted state. If the self-test passes, the FAILURE pin is set to the unasserted state; otherwise, the FAILURE pin is left asserted, indicating failure of the self-test. The self-test provides confidence that all major internal paths, the microcode ROM, and the registers are operational. More detailed tests can then be run that assume this basic functionality. The self-test also reads several locations in external memory to check that the data path and bus interface circuitry work correctly. This mechanism plays a key role in the methodology for testing in a system environment. For example, the system might connect the FAILURE pin to a LED on the processor board to visually indicate a failed board. The self-test covers more than 70% of the transistors.

- *Microcode patch.* Two microcode patch registers were provided to fix microcode bugs. This mechanism was used to test revisions of the microcode prior to committing it to ROM as well as to fix microcode bugs in early steppings.

- *Processor logout function.* The logout IAC function caused the processor to write its internal state into external memory. The state information included all register values and the contents of the instruction cache and the TLB. This IAC message existed in the early steppings but was removed in the final stepping.

- *Debug logic.* Three pins control some logic to drive and observe the microinstruction bus and two of the internal data buses. This mechanism helped test and debug early silicon. These pins are not accessible on production parts.

THE MICROARCHITECTURE

The following sections describe in greater detail the structures and implementation techniques used in the first implementation of the 80960 architecture.

Instruction Fetch Unit

The instruction fetch unit is responsible for fetching instructions and maintaining a cache of available instructions for the instruction decoder to process. It contains a direct mapped, 512-byte instruction cache, fetch and prefetch controllers for the cache, cache-miss processing logic, the instruction pointers and associated control logic, instruction breakpoint registers, test code support logic, and special fault support.

The instruction cache (Figure 6-3) is physically implemented as a 64-bit by 64-entry RAM array containing instructions and as a 48-bit by 16-entry RAM array for the address tag array. Logically, the former is organized as one hundred twenty eight 32-bit words of direct mapped instructions. The block size of the cache is 16 bytes (or four words). Thus, there are thirty two 16-byte blocks. The address tag array is logically 32 24-bit words. This cache organization was chosen based on die size considerations in conjunction with simulation results and data from the literature.[1-4] Other mapping schemes were evaluated (two- and four-way set associative) but were discarded because simulation results showed a net performance decrease for this size cache. This result may seem counterintuitive. In fact, making the cache mapping set associative does improve the logical performance of the cache when viewed on a cycle saving basis (for this size cache, the improvement is about 1-2%), but the circuit speed of such a cache is compromised because of the extra multiplexing logic at the output of the cache (one entry needs to be chosen from multiple possible entries that are accessed simultaneously). This circuit was a critical path that would have limited the operating frequency of the chip, and by removing this last stage, the cache access time was no longer the most critical circuit. Another significant factor swaying the decision was that a direct mapped cache is much simpler to implement and takes much less space, for example, only one set of sense amplifiers is needed. The 16-byte block size was chosen because our simulations and the literature show that this size is optimal for a 512-byte cache.

The instruction fetch unit also contains a set of six instruction pointers (IPs) to preserve the machine state for the pipeline stages and fault cases. These IPs are defined as follows:

1. The *fetch IP* is used to specify the word address of the next instruction to be accessed from the cache, or in the case of a miss, it is used to specify the word address of the target instruction in memory.

2. The *prefetch IP* specifies the quad-word address of the next block beyond the block that contains instructions currently being fetched using the fetch IP. The prefetch IP is used to fetch new blocks of instructions in advance of the instruction currently being fetched from the cache.

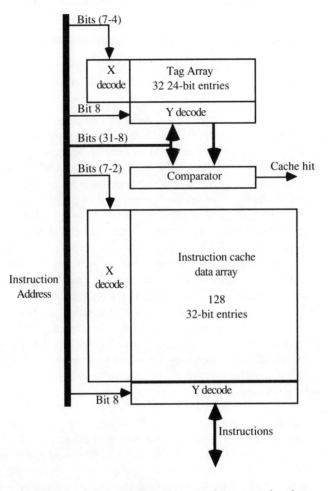

Bits (7-4)

X decode

Tag Array
32 24-bit entries

Bit 8

Y decode

Bits (31-8)

Bits (7-2)

Comparator

Cache hit

Instruction Address

X decode

Instruction cache
data array

128
32-bit entries

Y decode

Bit 8

Instructions

Figure 6-3:   Instruction cache organization

3. The *current IP* specifies the word address of the current instruction that is being decoded by the instruction decoder in the case of instruction execution or by the microinstruction sequencer in the case of microinstruction execution.

4. The *spin IP* specifies the word address of a second buffer in the instruction decoder for the case where the instruction decoder is spinning on its

currently executing instruction and the instruction fetch unit has fetched ahead another instruction.

5. The *buffer IP* and

6. The *execute IP* are saved values of the current IP which track the pipeline at various stages. They are used to recover from fault conditions.

Two 31-bit breakpoint registers are also included in the instruction fetch unit. A breakpoint is signaled if the upper 30 bits of either of these registers are equal to the current IP and if the low order bit, the breakpoint enable bit, is set.

The IPs, the breakpoint registers, a 32-bit displacement adder, and associated control logic are implemented in a data path separate from the integer execution data path. This allows all the IPs to be updated concurrently with other operations in the chip. This ability is critical if one instruction per cycle is to be executed.

The fetch and prefetch controllers are finite state machines implemented as PLAs (programmable logic arrays). They are responsible for keeping the instruction cache filled with the instructions that are the most likely to be executed. The fetch controller uses the fetch IP to fetch new instructions from memory whenever a miss occurs. The prefetch controller is responsible for trying to anticipate which block of instructions will be needed beyond the currently executing block. Whenever the cache is not busy (e.g., if the cache gets ahead of the decoder), the prefetch controller checks the cache directory to see if the next sequential block is already in the cache. If it is not, the prefetch controller uses the prefetch IP to fetch that block. The prefetch controller also fetches one block ahead whenever a branch is taken in order to fill the pipeline as fast as it can.

A special feature of the instruction cache is its ability to hold either instructions or test instructions. Control logic that determines the instruction type is used to gate the cache output to the instruction decoder in the case of instructions or to the microinstruction sequencer for direct execution. The latter mode of execution was extremely valuable as a chip debug tool and for test program development. Test code can directly control on-chip resources that would normally require long programs to achieve similar on-chip state conditions.

An example is testing the CAM (content addressable memory) part of the TLB in the address translation unit. Without the test-code capability, one must first cause a TLB miss to fill each TLB line with a unique value, and then perform a memory cycle that hit each successive line, ensuring that one and only one line in the TLB causes the hit. This procedure must then be repeated for the complement values in each CAM register. Doing all of this through the normal flow of execution is certainly possible but is difficult and requires many cycles (a TLB miss must be processed each time a new value is placed in the TLB). Given the ability to use test code from external memory, it is possible to directly load and compare values in the TLB CAM registers. Fault routines that are normally performed by microcode can also be bypassed, making the test much more straightforward and requiring significantly less execution time.

The instruction fetch unit also contains logic to test whether a cache block that causes a fault in the address translation unit (e.g., page fault) is really needed for execution. This check then prevents page faults from occurring on instructions that are not executed.

Instruction Decoder

The instruction decoder is the main control unit for the chip. It accepts instructions from the instruction fetch unit's instruction cache, decodes them, and either executes them directly, issues a microcode flow, or vectors control to the microcode ROM. In the case of branch instructions, the instruction decoder contains all of the necessary logic to execute the instructions directly. With this logic it is possible for a branch instruction to be executed in parallel with other instructions (e.g., a multiply in the floating point unit or a multiword move in the integer execution unit). Thus, as will be described in Chapter 8, branch instructions can sometimes consume zero cycles for execution. For most instructions (other than branches), the Instruction Decoder transforms the instruction into a similar microinstruction, which is then gated directly to the microinstruction bus. For the more complex instructions, the instruction decoder vectors to the microcode ROM and relinquishes control until the ROM sequence is complete.

An important role of the instruction decoder is providing a scoreboard for the condition code. This mechanism minimizes the impact of branching on pipeline performance. In a pipelined microarchitecture without condition code scoreboarding, a conditional branch must wait until all instructions in the pipeline have finished executing because one or more of them could alter the condition code. Any instruction that alters the condition code marks the condition code as busy until the instruction finishes. (Significantly, in the 80960 architecture, only a limited number of instructions, the compare instructions, alters the condition code). Any instruction that uses the condition code (primarily conditional-branch instructions) waits if the condition code is busy; otherwise it is executed immediately. The use of this technique by software is shown in the following example:

```
(A)                      (B)
addo    r2,r3,r4         cmpo    r2,0
cmpo    r2,0             addo    r2,r3,r4
be      xyz             be      xyz
```

In sequence A, we are adding r2 and r3, comparing r2 to 0, and then branching if the condition code (set by the compare) indicates equality. Although the branch is seen by the decoder in parallel with the execution of the comparison instruction, the processor must wait for the comparison to set the condition code before deciding whether to branch. In the equivalent sequence B, the code is rearranged, taking advantage of the existence of an instruction (the ADDO) that neither alters nor uses the condition code, and thus can be moved between the compare and the branch. Sequence B is logically equivalent but executes in less time. This reordering of code is a simple optimization for compilers to perform.

The instruction decoder is organized as a set of decoding and sequencing control PLAs. One PLA is used to decode the simple instructions that are directly executed by the instruction decoder; this PLA also checks for invalid addressing modes and invalid opcodes. Control signals from this PLA gate the instruction fields onto their respective positions on the microinstruction bus. Another PLA controls the sequencing of the instructions that require a short sequence of microcode to execute. This PLA gates onto the microinstruction bus a canned sequence of instructions performing simple operations such as multiword load/store instructions, effective address calculation, or floating point instruction flow. An example of an effective address calculation would be to scale an index and add that to a displacement value. The floating point instructions require the instruction decoder to move source operands from the register file in the integer execution unit to the temporary registers in the floating point unit, issue the appropriate execute microinstruction, and move the result back to the destination register.[*] Another set of decoding logic provides a vector to starting addresses in the main microcode ROM. This vector is provided by bit steering the opcode field of the instruction to a subfield for the next microcode address, which is then used as an index into the branch table for the starting addresses of all of the microcode routines.

The instruction decoder also provides the synchronization logic necessary to allow any tracing parameters that are set to divert the machine state to allow microcode to handle any trace events. Whenever a trace event is enabled and the relevant condition is signaled, an individual trace-event signal is written into an appropriate bit in a control register for later interrogation by microcode. At the end of the instruction the event is signaled to microcode as a microinterrupt.

Finally, as mentioned above, the instruction decoder provides all of the execution logic for the branch instructions. This includes computing a target displacement and adjusting the appropriate IPs in the instruction fetch unit (namely the current and fetch IP). For conditional branch instructions, the same occurs but the branch is gated by the result of the condition code and the condition specified in the instruction. Compare-and-branch instructions are also handled directly by the instruction decoder and sequenced as they would be if they were individual instructions.

---

[*]Floating point instructions that specify an 80-bit floating point register as an operand do not require this sequence.

Integer Execution Unit

The integer execution unit provides the core resources and control mechanisms through which most instructions are executed including all the simple arithmetic, logical, and bit instructions. Operands are fetched from the register file or forced to a constant value if a literal is specified. The main control is decoded from the microinstruction bus. Some local sequencing is done directly by the integer execution unit for certain basic functions (e.g., data alignment, multiword register moves, and memory operations). The integer execution unit also sends data to and from the address translation unit for memory operations and to/from the floating point unit for floating point operations. It signals local register management microassists and integer overflow faults to the microinstruction sequencer for microcode fault handling.

Figure 6-4 shows the main components of the integer execution unit data path. The register file is integral to this unit. All data that is passed between memory and the processor is passed through the register file, which is an array of 32-bit entries and includes the global registers (excluding the 80-bit floating point registers), 4 sets of local registers, and 34 scratch registers. Thirty-two of the registers (the global registers and one of the local register sets) provide the programming environment required by the architecture. The other three local register sets provide an on-chip stack of register sets. This greatly improves the performance of procedure call and return. The scratch registers are used by microcode for temporary storage and for computing intermediate results. These registers are not visible to the programmer.

Due to the large number of registers and the die size constraints, the register file is single ported (the standard six-transistor RAM cell was used). To overcome the possible performance limitations of the single-ported array, the array was designed to be able to read one register and write another register in the same cycle, the read occurring in the first phase[*] and the write occurring in the second. Also, a final pipeline stage was added to allow the deferred write back of instruction n to be concurrent with the execution of instruction $n+1$. This technique required the addition of a bypass path that allows the result of instruction n to be gated to the input of the execution data path, instead of the register addressed in instruction $n+1$. This bypass selection would occur if the address of the destination register of instruction n were the same as a source register of instruction $n+1$. These techniques allow many instruction sequences to be performed at a net rate of one instruction per cycle.

The remainder of the integer execution unit data path is comprised of a 32-bit arithmetic logic unit (ALU), the condition code flags, arithmetic flags, a 32-bit barrel shifter, sign extension logic, the register file bypass path described above, and local register allocation logic.

---

[*]The internal clock cycle is divided into two equal phases.

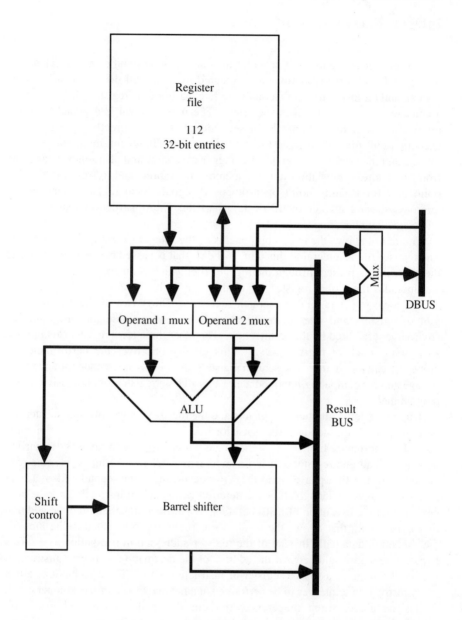

Figure 6-4:   Integer execution unit data path organization

The ALU is a 32-bit dynamic circuit that can perform any of its operations in one clock cycle. It supports the following functions: add, subtract, and compare on

32-bit integer and ordinal values, and the 16 bitwise-logical operations on two 32-bit ordinal values.

The output of the ALU is combined to form the three condition code flags (greater than, less than, and equal), which are loaded into the arithmetic control register whenever an instruction is executed that alters the condition code.

Several other flags in the integer execution unit assist microcode with some of the more complex instructions. Examples of these are the ALU result sign bit, ALU result = 0, ALU low order result byte = 0, and ALU result greater than 9 (decimal carry).

The barrel shifter performs bit shifting from 1 to 31 bits in one clock cycle. It supports all of the shift, rotate, and bit instructions and is used by the integer execution unit sequencer to do byte alignment of data. When doing alignment operations, the barrel shifter also is capable of sign or zero extending the result for values that are less than 32 bits in length.

The local register set allocation logic ensures that when a procedure call or return instruction is executed, the new set being used is a valid one. It does so by keeping a set of valid bits for each set and a pointer to the current set. When a call is executed, the pointer is incremented and the valid bit for the new set is checked. If it is already valid, the registers have data that must be flushed to memory before the new set can be used, in which case control is vectored to a flush register set microassist prior to completing the call. If it is not already valid, that is, the register set is available, the valid bit is set and the call is completed. When a return instruction is executed, the current register set valid bit is reset and the pointer is decremented. If the new set (the one associated with the procedure to which the program is returning) is not valid, it needs to be restored from memory before the return can be completed. In this case control is vectored to a restore register set microassist prior to completing the return. Otherwise, the new set managed to stay on chip and the return is completed. The most common case for both the call and return is that the register set is available and no memory references are required to save or restore registers across the procedure boundary. Note that the above process is entirely transparent to the programmer.

All the control for the integer execution unit is provided by the integer execution unit sequencer, which contains a decoding and sequencing PLA, register address latches, shift count control, and multibyte move logic. The PLA decodes microinstruction opcodes and asserts the appropriate control signals for the functional blocks in the integer execution unit data path. It also sequences some multicycle instructions requiring local decisions that affect the operation of subsequent cycles. An example of this is the shift right dividing integer instruction, which requires knowledge of the shifted result to compute a possible correction cycle. The register address latches control the register file. The shift count control logic provides the shift count and shift direction control for the barrel shifter. It also controls which operands form the input to the barrel shifter and whether to sign extend the result.

The multibyte move logic keeps track of the byte position and length of unaligned multibyte data transfers.

Bus Control Logic

The bus control logic provides the interface between the internal chip logic and the local processor bus. Two major data paths exist between the bus control logic and the other units. The DBUS is used to transfer information between the bus control logic and the integer execution unit, and the BIBUS is used for address transfer from the address translation unit and instruction transfer to the instruction fetch unit. Since the BIBUS can be used to transfer both an address and an instruction in any cycle (i.e., it is driven in both phases of the clock), there is no need to arbitrate for this bus. Data returned to the register file requires the bus control logic to arbitrate for the microinstruction bus and assert a "move_from_BCL" microinstruction to make the transfer.

The bus control logic contains an output FIFO, an input FIFO (see Figure 6-2), a controller for interrupts and IACs, the local processor bus state sequencer, and some debug logic.

The output FIFO buffers outbound requests that are waiting for access to the local bus. The input FIFO buffers inbound data that is waiting for an available cycle to transfer information to a destination register. The FIFOs support up to three outstanding memory accesses. Each of these accesses can contain 1 to 16 bytes of data. The performance effect of these FIFOs is significant because the internal execution is not delayed by the bus until either an instruction needs data from memory or the output FIFO is full. The first case is synchronized by the scoreboard logic as described in the preceding sections. The latter case is rare.

The interrupt control logic supports the four interrupt pins and the IAC signal. The interrupt pins can be used as four independent interrupts or as two interrupts and an interrupt request/acknowledge pair for compatibility with an 8259A interrupt controller. This allows expansion of interrupts if more than four are needed.

The local processor bus sequencer is implemented as a finite state machine that tracks the state of the external bus and provides the signals that control data movement between the FIFOs and the bus control logic. A separate PLA exists to compute the values for the byte enable pins.

Finally, the bus control logic contains some debug logic that places the chip in a mode such that the microinstruction bus or the DBUS can be driven or observed from the external bus. This feature was very useful in debugging initial silicon, but is not provided to users.

Microinstruction Sequencer

The microinstruction sequencer is responsible for all control flow that requires sequencing that is too complex to be handled by the other units. Examples of such flow include fault conditions, microassists, interrupts, initialization, self-test, and complex microcode routines.

The microinstruction sequencer contains the microprogram ROM, microaddress logic, microinstruction format control, several control registers, the register scoreboard, and the interrupt control logic.

The microprogram ROM is organized as 3072 42-bit words. Each microinstruction contains two parts: the normal instruction that is emitted on the microinstruction bus and address information for the next microaddress. Microinstructions have one of four possible formats; the microinstruction format logic decodes and gates a single format onto the microinstruction bus and the microaddress register.

The microaddress logic contains a microprocedure stack, two patch registers, a multiway branch register, and the microaddress register. The microaddress stack is provided to allow microprocedure calls up to eight levels. The two patch registers allow two ROM words to be patched by forcing a branch to external memory when one of the patch addresses is accessed. An external program can then fix the problem and return control to the microcode. The multiway branch register provides a branch mechanism by computing the logical OR of the next address information from the current microinstruction with a register that is presettable by microcode. These functions together provide a very flexible microcode flow control mechanism. Some of the possible control alternatives are

- conditional branch on 1 of 32 flags

- conditional delayed branch on 1 of 32 flags (i.e., execute the current microinstruction while waiting for the condition to settle and then take the branch)

- conditional branch if loop counter (described below) is not zero, and decrement the loop counter

- conditional call/return on 1 of 32 flags

- multiway branch based on contents of multiway branch register

Control is also provided to manipulate the microaddress stack without affecting the next address. These functions include push, pop, and clear the microaddress stack.

Several registers in the microinstruction sequencer keep information associated with the program status and microcode state information. Examples are the process and processor status, trace control, and process timer registers. Other special function registers exist to aid microcode. Examples include the microcode loop counter, execute register, and multiway branch register. The loop counter, for instance, allows parameterized microcode loops for functions, like those required for the string instructions.

The microinstruction sequencer contains an *execute* register that allows microinstructions to be built and executed by other microinstructions. This provides an efficient mechanism for the parameterization of certain fields of a microinstruction. For example, the self-test microroutine cycles through all of the registers. It can do this by creating a microinstruction in the execute register to add a register to a checksum after it has been loaded with some data pattern. It can then increment the register specifier and execute the same instruction to add the next register. A loop can be set up to run through all of the registers with only a few words of code. The entire self-test microprogram requires less than 100 words and tests more than 70% of the transistors on the chip. The execute register and loop counter are the main contributors to the code compaction.

The interrupt control logic provides a central area for the handling of asynchronous events. It prioritizes all of the events and vectors microcode to the appropriate handling routine. The events include the following:

- A bad access is received from the bus interface

- The integer execution unit needs a microassist

- The floating point unit has detected a fault

- The floating point unit needs a microassist

- The address translation unit has detected a fault

- The address translation unit needs a microassist

- The process timer has reached zero

- One of the four interrupt pins is asserted

Address Translation Unit

The address translation unit provides the hardware necessary to implement efficient on-chip memory management. It performs virtual to physical address translation for all memory operations and initiates memory accesses to the bus control logic. A miss causes the address translation unit to send a virtual address microassist to the microcode so that a TLB entry can be fetched from memory.

Memory translation requests can originate from two sources: from the microinstruction bus in response to a request made by the instruction decoder or microcode, and from the instruction fetch unit in response to a fetch or prefetch for instructions to fill a block in the instruction cache. The first case is the normal request path for all data accesses. The second provides an independent and asynchronous path for instructions.

The address translation unit's other functions include

- verification of access rights to a page

- support for splitting an access of a multibyte request that spans a 16-byte block boundary into two separate contiguous requests

- replacement of address translation unit entries according to an algorithm that approximates the least recently used algorithm

- support for invalidation of TLB entries by microcode

- notification to microcode when the instruction needs to update the altered or accessed flags in a page table entry

- automatic lookup of page table entries on a page miss without microcode support

The address translation unit contains a TLB of 48 entries subdivided into the following categories: 32 general-purpose, 4 local register set and 12 dedicated entries. The 32 general-purpose entries are controlled by the replacement logic. They are subdivided into four groups of eight entries each. The replacement logic keeps track of the least recently used group. A pseudorandom entry* in the least recently used group is replaced on a miss. The 12 dedicated entries contain specific translation descriptors for memory locations that are used by microcode to guarantee addressibility at all times. Examples of these are the current process control block, the interrupt stack page, and the four regions of the current address space.

The virtual address buffer stores the virtual address that is being translated. It also contains logic to manipulate the offset in the event of a multibyte access that spans a 16-byte block boundary. In that case, the access is broken into two separate memory requests with the appropriate length and offset generated to keep the bus from accessing data across the block boundary. This feature provides a much simpler external bus interface than would otherwise be required if the bus supported multibyte access to arbitrary byte boundaries. For example, a 2-bit counter in external logic is required to track the multiword bus cycles, rather than the 30-bit counter that would be required if the access were aligned to an arbitrary boundary. Also, systems that implement multiple buses can accommodate mid-level interleaving for bus load balancing and do not require cooperation of the controllers that are performing the bus transactions. This is exactly the mechanism supported in the BXU chip.

---

*These pseudorandom numbers are generated from a modulo 8 counter that is incremented at the end of each instruction.

Floating Point Unit

The floating point unit provides full support for the floating point operations and the multiply and divide operations for integers and ordinals. It is a self-contained unit with a relatively simple interface to the rest of the chip. The floating point unit supports the full set of data types defined by the IEEE standard[5] (real, long-real, and extended-real) as well as the full complement of instructions defined in the numerics architecture. The floating point unit interacts with the integer execution unit for data transfers to and from the register file, and with the microinstruction sequencer for microcode assistance. The floating point unit performs most functions with its data paths and internal sequencer, but relies on microcode support for pre- and post-operation calculations, as well as for faults and microassists. Examples of these include normalization; rounding; and response to overflow, underflow, divide-by-zero, invalid-operand, real-arithmetic, and inexact faults.

Although the floating point unit was initially designed for the 80960KB, its simple interface and self-contained nature made it an ideal module to be used as the basis for the Intel 80387 coprocessor chip for the Intel 80386. Thus, both the 80960KB and 80387 provide floating point with identical hardware and inner loop algorithms. While the hardware is the same and the performance of the 80386/80387 is indeed impressive, the performance of the 80960KB exceeds that of the 80386/80387 pair by a significant factor; for example, the Whetstone benchmark for the 80960KB runs more than two times faster than that of the 80386/80387. This is primarily due to the coprocessor overhead associated with the transfer of data operands between the 80386 and the 80387.

Many tradeoffs had to be made in the floating point unit because of die size limitations. We budgeted one eighth of the total chip area to the floating point unit, which precluded the use of a multiplier array. The most feasible alternative was a single 68-bit adder with a trailing shifter. This combination, along with some skipahead logic, provided a much smaller block, which statistically approaches the performance of a multiplier array.

A second tradeoff was made in the shifter design. Normalization of floating point operands and results requires a variable-length shift of 1 to 66 bit positions. A shifter that large would consume twice the area of the 68-bit adder. The shifter that was implemented can shift 1 to 16 bit positions and requires 25% of the area required by a 66 bit shifter. Several passes are required in the worst case; however, the worst case is statistically rare and has been shown to occur less than 4% of the time.[6]

The basic algorithm in the floating point unit for multiply operations is a 2-bit-per-cycle modified Booth algorithm.[7] The design also employs skipahead logic if the next 3 or 7 bits are all zeroes or all ones; thus the multiply time is data-dependent. This has the largest benefit if one of the operands is a small number, as is often the case in integer multiply operations. With the skipahead logic, the best-case time for a 32-bit integer multiply is 7 cycles.

The transcendental operations are implemented primarily in PLA and ROM code, along with a small shifter and some local ROM constants. The CORDIC (coordinate rotational digital computation) algorithms[8] are used for the transcendental functions. The floating point unit's local sequencer provides the inner loop control for the main algorithms (pseudomultiplies and pseudodivides). The total cost of these operations is about 8% of the area of the floating point unit. This includes 350 words of main microcode, 25 terms in the floating point unit sequencer PLA, a 16-bit shifter, and 32 words of the mantissa ROM.

Finally, the floating point unit implements a separate data path for exponent and mantissa calculations. This allows fully parallel manipulation of these logically separate parts, resulting in a 50% speed improvement over that obtained with a single data path for add and subtract operations, the most frequent floating point operations. The cost of the separate 16-bit data path and adder is about the same as that of the 68-bit adder. Performance numbers for the floating point operations can be found in Chapter 8.

The floating point unit consists of a sequencer, a 68-bit mantissa data path (Figure 6-5), a 16-bit exponent data path (Figure 6-6), the four floating point global registers, and an operand interface.

The operand interface provides the data path and multiplexing logic to move data between the integer execution unit register file and the 80-bit floating point registers. The data is transferred in 32-bit words and is packed and unpacked from and to the internal 80-bit format. This interface also recognizes special cases of floating point operands during the transfer process (e.g., invalid and unnormalized operands), which are then used to gate the sequencer or cause a microassist or fault.

The sequencer provides the inner loop algorithms for all the operations supported by the floating point unit. The sequencer provides control to the hardware blocks until such time as a fault, microassist, or completion is signaled to the microinstruction sequencer or until the result is returned to the destination register. During any sequence, the floating point unit operates in parallel with other activity in the chip. Thus, the floating point unit provides fully autonomous operation as long as subsequent instructions do not need its resources.

The mantissa data path (Figure 6-5) is 68 bits wide (mantissa plus the guard, round, and sticky flags) and consists of a shifter, adder, 44-word ROM, and three temporary registers (two operand registers and an accumulator). The ROM provides constant values that are used in the CORDIC algorithms for the transcendental operations, rational approximation constants, and other, miscellaneous values. The adder can perform a 68-bit add/subtract with some input scaling in a single clock cycle. The accumulator latches the result (scaled by 0, 1, 2, 4, or 8 bits). The shifter takes a 68-bit operand and shifts right or left by 0-16 bits per cycle. The guard, round, and sticky flags are required for computing a rounded result and to determine if any precision is lost.

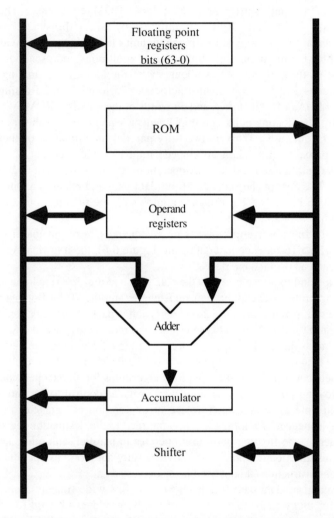

Figure 6-5:   68-bit floating point mantissa data path

The exponent data path (Figure 6-6) is 16 bits wide and consists of an adder, a 4-word ROM, two operand registers, and some sequence detection logic. The ROM contains the exponent bias values. The detection logic checks for overflow and underflow faults, and correction cycles.

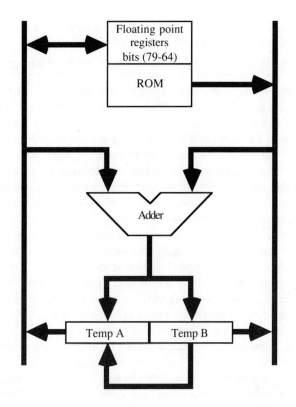

Figure 6-6:   16-bit floating point exponent data path

The four floating point global registers that are visible to the instruction set as four global registers are implemented as part of the floating point unit data path. Each register consists of a 68-bit mantissa, as part of the mantissa data path, and a 16-bit exponent, as part of the exponent data path. The literal generation logic is also part of this block.

A LOOK TO THE FUTURE

Although this chapter has mainly dealt with the first implementation, it is worthwhile spending some time discussing the possibilities for future implementations. As is traditional with most microprocessor families, many different versions of successful product families are ultimately built. The 80960 architecture lends considerable flexibility in this regard, since the architecture was developed with future implementations in mind. The multiple architecture levels, for instance, allow one to construct various implementations to meet different market needs. For example, a highly integrated microcontroller based solely on the core architecture level could provide all of the core instruction set, some typical I/O peripherals (such as serial ports, parallel ports and timers), some on-chip ROM or EPROM, and some on-chip RAM. This would provide a very low cost chip optimized to meet the needs of a well defined market with full instruction set compatibility of a predefined subset of the architecture. Similar scenarios exist for the other levels of the architecture. A major consideration for system and software developers is that the first implementation provides a vehicle compatible with future subset implementations. The compilers and software development tools were also designed with future implementations in mind.

The preliminary design of such a microcontroller that is a follow-on to the 80960KA is currently under development at Intel. This design is an implementation of the core architecture, with many high-integration features. This chip represents an exciting development in performance for low-cost, high-integration microcontrollers. It contains many performance enhancements, such as parallel instruction execution, a target operating frequency of 32 MHz, and very high bandwidth buses (more than 500 MB/sec) to internal memory.

This chip is capable of executing a branch instruction, an arithmetic instruction and a memory operation concurrently. Special instruction decoding logic is employed to look at the next four instructions in parallel. These techniques have made it possible to achieve sustained instruction execution speeds of 64 MIPS, or two instructions per cycle.

REFERENCES

1.   A. J. Smith, "Cache Evaluation and the Impact of Workload Choice,"
        Proc. 12th Annual Symp. on Computer Architecture, ACM, 1985, pp. 64-73.

2.   A. J. Smith, "Sequential Program Prefetching in Memory Hierarchies,"
        Computer, Vol. 11, No. 12, 1978, pp. 7-21.

3.   J. E. Smith and J. R. Goodman, "Instruction Cache Replacement Policies and Organizations," *IEEE Trans. on Computers,* Vol. C-34, No. 3, 1985, pp. 234-241.

4.   A. J. Smith, "Cache Memories," *Computing Surveys of the ACM*, Vol. 14, No. 3, Sept. 1982.

5.   ANSI/IEEE Standard 754-1985: "IEEE Standard for Binary Floating Point Arithmetic," *IEEE*, 1985

6.   D. W. Sweeney, "An Analysis of Floating Point Addition," *IBM System Journal* , No. 4, 1965, pp. 31-42.

7.   Andrew D. Booth, "A Signed Binary Multiplication Technique," *Q. J. Mech. Appl. Math.*, No. 4, 1951, pp. 236-240.

8.   Jack E. Volder, "The CORDIC Trigonometric Technique," *IRE Transactions on Electronic Computing*, Vol. EC-8, 1959, pp. 330-334.

CHAPTER 7

Design Methodology

Although, in most respects, the tools and methodologies used to create the 80960 product line may be viewed as irrelevant to understanding and using the 80960, a discussion of the development process gives additional insight into the product. This chapter summarizes the development process and shows the interaction that occurred between the tools and methodologies on one hand and the product definition on the other. The tools and methodologies used in the development of the 80960 affected its definition in indirect ways by instilling confidence in the ability to produce a product at the 80960's high level of function, integration, complexity, and performance. Conversely, the definition of the 80960 had significant influence on the characteristics of some of the tools and methodologies used.

ARCHITECTURE METHODOLOGY

The development of the architecture was driven from three points of view. The first was a set of market requirements for embedded microprocessor applications. This led to such features as integrated floating point and the 80960MC's tasking and protection support for the Ada environment. The second was the desire to innovate, which resulted in such things as the priority interrupt mechanism, mixed-length floating point operations, and the tracing facilities. The third was the performance targets, which influenced the basic instruction-set design, the register model, the scoreboards, the instruction cache, and many other aspects of the product.

Several tools were developed that had a major benefit to the architecture process, one being a *macrosimulator* of the entire processor architecture. The macrosimulator, a C program, implemented all aspects of all three architecture levels, including all implementation-dependent aspects. As such, the macrosimulator could "execute" any 80960 program. Rather than serve as just a bare machine, the macrosimulator also contained features such as single-stepping, tracing, and display and modification of memory.

The original purpose of the macrosimulator was performance modeling. Although it did not mimic the actual chip implementation identically (in terms of performance), it did accurately model the major functions in which we were interested, such as the instruction cache. Because the macrosimulator ran as an application program atop a mainframe running UNIX, a clever feature incorporated in the macrosimulator was a special 80960 pseudo-instruction representing a UNIX system call, which told the macrosimulator to escape to the underlying UNIX system to perform a UNIX function (e.g., input/output) for the 80960 program under execution. This coupled with the C compiler discussed later gave us a large source of real programs (UNIX applications and utilities) to use for performance analysis. In addition, benchmark programs were developed by the project members, among them Dhrystone.[1] The Dhrystone benchmark has since become one of the best-known industry-wide benchmarks.

Although the original purpose of the macrosimulator was performance analysis, its existence as an independent (and early) implementation of the processor architecture gave it several other uses. It was used extensively by software developers for debugging of 80960 software prior to the arrival of the A-step* processor chips. The macrosimulator was also used as a vehicle for debugging architecture test suites.

Three other tools were developed at the same time as the macrosimulator: an assembler, C compiler, and image builder. The principal reason for developing an early "throwaway" C compiler was hinted at above: it created access to an enormous base of programs for performance analysis. The image builder was a tool that constructed memory images consisting of such entities as programs, page tables, and process control blocks.

Another simulation tool developed early in the project was the *interconnect simulator*, which was able to model arbitrary system configurations consisting of processors, BXUs, AP buses, and caches. Its stimulus was memory references from processors, and this stimulus could come from built-in memory-reference generators based on specified probability distributions or from trace files constructed by the macrosimulator. The interconnect simulator was used for system performance analysis.

---

*The term *step* or *stepping* denotes the fabrication of wafers for a complete version of a chip. The A-step is the very first version of the chip, the B-step the first complete revision, and so on.

A major part of the architecture effort, in terms of cost, was the construction of an extensive set of architecture test suites, intended to determine the conformance of an implementation of the architecture to the architecture specification. These test suites were partitioned by each of the architecture levels, and can be used on future implementations of the architecture. These test suites (as opposed to other test suites discussed later) view the implementation as a black box and are composed of four categories of tests:

1. Machine-generated tests
2. Hand-crafted tests
3. A multiple-process system test
4. Special floating-point tests

For the first category, a LISP program named GT (generator of tests) was developed which, given the semantic description of an instruction and the range of possible input values, would generate a program in 80960 assembly language to exercise all "interesting" usages of the instruction. All of these tests (as well as all of the tests in the other categories) were generated to be self-checking. In other words, each test compares the expected result to the actual result. This self-checking extends not only to the obvious, such as the output value and condition code, but also to things such as faults. For instance, when GT generates a test for a division instruction using a zero-valued divisor, it also generates code to determine whether the divide-by-zero fault actually occurred.

Although GT successfully produced all test suites for the basic instruction set, it was not capable of producing tests for complex functions (e.g., process dispatching, page faults, interrupts) or for situations where human ingenuity is needed. Therefore a significant amount of time was spent producing hand-crafted test suites. These followed the same structure as the GT-produced test suites (e.g., they were self-checking).

A weakness in these two categories is that they focus on testing individual functions, but not on the complex interactions *among* functions. To compensate for this, several versions of a large system-test program were written. Because most of the complex interactions occur in the protected architecture, the system test was written for the 80960MC. Because the 80960MC is a superset of the 80960KA and 80960KB, and because all three products were derived from a single design, incorrect interactions in the 80960KA and 80960KB would be discovered by the 80960MC system test.

The system test consists of multiple self-checking parallel processes, with considerable randomization of what they do and how they interact. For instance, one process creates page-fault situations in other processes by choosing random page-table entries at random times and marking them as invalid. This process also randomly changes the priorities and time-slice values of the other processes, and the resultant behavior is tracked by watching the execution time of the other processes.

Other processes in the system test interact with each other by sending messages through ports and by signaling semaphores. Finally, the system-test program contains a mechanism to inject interrupts during the test.

The final category of architecture tests is floating point. In this category four approaches were used. First, an existing test suite for the operations in the IEEE floating-point standard was used. Second, hand-crafted tests were developed for situations not covered by this test suite (e.g., faults) and for 80960 functions that extend beyond the standard. Third, an automated test generator was developed, principally for the transcendental instructions. Given a set of interesting data points and an error bound, the automated test generator computes the infinitely precise result expected and determines if the actual 80960 result is sufficiently close to the infinitely precise result.

The fourth category of floating-point validation required a different approach. The architecture requires that functions be *monotonic*; for instance, in the range $0..\pi/2$, $\sin(x)$ must be less than or equal to $\sin(x+\varepsilon)$. Determining monotonic behavior by test-case execution for all possible values is infeasible because of the execution time required. Therefore, a different approach was used; mathematical analysis of the floating-point logic and microcode was done to prove monotonicity.

Except for the monotonicity analysis, all the testing described above is black-box testing. A weakness of black-box testing is that it does not use the "clues" to useful test cases that can be found by examining the insides of the processor. Since most of the processor's complex functions are implemented in microcode logic, there is a lot to be gained by tailoring test cases to the microcode logic. The architecture test suite was supplemented by knowledge of the microcode design, as discussed in a subsequent section.

CHIP-DEVELOPMENT METHODOLOGY

The methodology employed for the processor development is illustrated by the nine steps in Figure 7-1. The major step, by far, is the development of a microsimulator, a software representation of the internal behavior of the processor. The microsimulator, or RTL (register-transfer-level), model is a complete, clock-phase by clock-phase software simulator of the chip at a level of detail sufficient to allow design engineers to translate the microsimulator code directly into logic schematics. The reason that so much emphasis is placed on the microsimulator is that once it is verified as correct, it serves as the standard against which all other levels are validated.[2]

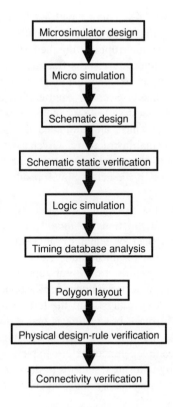

Figure 7-1: Major processor design steps

    The level of detail modeled in the microsimulator varies from section to section;
the only criterion used is that relatively inexperienced engineers must be able to
take a page of program code from the microsimulator and derive a schematic circuit
diagram from it. Therefore, things like internal buses, registers, PLAs, and random
control logic are expressed explicitly in the microsimulator. The major difference
between the microsimulator and the macrosimulator is that the microsimulator is an
exact, clock-phase by clock-phase representation of the processor logic, and as such
runs 80960 programs about two orders of magnitude more slowly than the
macrosimulator.
    The speed of the microsimulator, however, was still orders of magnitude faster
than that of a gate-level simulator, and much emphasis is placed on structuring the
microsimulator so that a compile-link-debug sequence occurs in a matter of min-
utes. The speed of simulation and speed of correcting mistakes had a significant

effect on the productivity of the design engineers. On a VAX-11/780 computer, the microsimulator simulates about eight 80960 cycles per second, and, prior to the A-step tapeout,* over 10 million cycles of test programs were run on the simulator.

The microsimulator also has an intimate relationship with the succeeding steps, and as a result the microsimulator is always an exact representation of the chip. For instance, if an error in a schematic is found in a subsequent step and the error also exists in the microsimulator, the microsimulator is updated prior to changing the schematic.

The microsimulator was validated in three ways. First, all of the architecture test suites were run on it. Second, engineers who developed different blocks of the microsimulator developed additional test suites tailored to the logic of that block. Third, the microsimulator was used for all microcode debugging and testing as described in the next section.

The next step was the translation of each section of the microsimulator into logic schematics. Three types of validation were done with the schematics. The first was done with a CAD tool that examined the schematic netlist with respect to 19 design rules. Examples of these rules are

- ensuring that each dynamic latch is refreshed in every cycle
- ensuring that there is at least one stage of static logic between latches that are used in different clock phases
- ensuring that the outputs of dynamic circuits are trapped
- checking for Miller capacitance coupling

The step after this was logic simulation. Since full-chip logic simulation requires vast amounts of computing power, logic simulation was primarily done on a block-by-block basis (examples of blocks are the integer execution unit, instruction decoder, and floating-point unit), and full-chip simulation was done only to verify interconnections among blocks. Since the schematics are just a lower-level representation of the microsimulator, all signals should match on each clock-phase boundary, and the correctness of the translation of the microsimulator to schematics is done by "playing" them side by side. Approximately 350,000 vectors (test cycles) were used to validate the chip at the logic-simulation level. Of these, only 22,000 were run on the full-chip model.

Two sets of tests were developed to validate the schematic-level representation of the chip: the design-validation (DV) and product-validation (PV) tests. The DV tests were oriented toward getting the quickest validation of a sheet of schematics; shortcuts were taken where appropriate. There was no requirement, for instance, to stimulate the circuit under test from the chip pins, nor was there a requirement to

---

*A *tapeout* is the event when all chip design work is complete and the database containing the chip layout is transmitted to the organization responsible for the production of masks.

observe the response at the chip pins. Also, critical circuit speed paths were ignored. These tests attempted to cover 90% of the transistors and were not intended as a production test. They were, however, expected to be used as the basis for the PV tests. And finally, these tests were required to be completed before the A-step tapeout.

The PV tests extended the DV tests to include stimulation from the pins and response observable at the pins. These were to be the final tests that are used on the production test floor for product shipment. Test vectors for critical circuit speed paths were also developed for the PV test suite. And finally, 100% coverage was the goal.

This development methodology provided the fastest possible validation of the schematics prior to tapeout. The PV tests did not need to be ready until six months later. The final PV test suite resulted in 350,000 vectors. Fault grading on a special simulation computer later determined that the initial PV tests covered 97% of all possible faults.

The third type of validation performed on the schematics was a new type of timing analysis. A database was maintained of timing and load-capacitance information of all significant signals (mainly signals interconnecting major blocks of the chip). Analysis of this database was then performed to find critical paths, which were then subjected to further circuit simulation. This analysis played a big role in avoiding the experience of many previous microprocessor projects, where several silicon steppings were required to get the designs to their clock-frequency goals.

The next step in the process was the physical layout of the chip, or the construction of the database that is eventually used to produce masks. To pack the large amount of function of the processor into a manufacturable die size, the layout was done without the aid of any automatic place-and-route tools. Two verification facilities were used on the layout database. First, a physical design-rule checker was used to ensure that the layout corresponded to the design rules of the specific CMOS process used for fabrication. The second was a proprietary CAD tool that provides connectivity verification between the layout and schematics (i.e., ensures that the layout matches the schematics).

Chip debug was performed on several benchtop VLSI testers, using the microsimulator to produce cycle-by-cycle test vectors. Since debugging the initial wafers of a chip of this complexity could have been a Herculean task, several key aids were designed into the chip. Several debug pads, which were bonded to unused pins in test chips, allow one to monitor and drive key internal buses from the tester. Also, a mode of operation was designed into the chip to allow microinstructions to be fetched from the external bus instead of the internal ROM. This allowed engineers to write DV tests that more readily access a particular internal unit than what could be done from machine instructions.

Another investment was made with chip debugging in mind; a voltage contrast capability was added to a scanning electron microscope to allow engineers to analyze internal AC signals on the chip without imposing the capacitive load of a probe. Although a substantial investment was made in developing this SEM capability, it was never used for debugging because the chip worked well enough from the beginning.

Production testing was done with a large-scale VLSI tester. Many of the production (PV) tests use the external microcode capability to enhance their fault coverage. Also, the chip contains the ability to "patch" internal microcode. Two address registers are present, and an attempt to fetch a ROM microinstruction from one of these addresses causes the chip to fetch microcode externally from a predefined memory address. The PV tests use this patch capability to shorten the number of vectors needed for certain fault-detection tests (e.g., tests of the address-translation errors use this to bypass the fault-handling microcode that would normally get invoked). This patch capability also served another important role: to patch serious errors in the microcode until the next stepping of the chip was available.

MICROCODE-DEVELOPMENT METHODOLOGY

Although the majority of the instruction set is executed without microcode interpretation, the processor contains a large microcode ROM for such functions as floating-point algorithms and special-case handling, string instructions, IAC message processing, and the process-management functions of the protected-architecture level.

The microcode was developed using standard software-engineering practices. All debugging and testing of the microcode were performed on the processor microsimulator, which had two benefits: (1) assurance against last-minute surprises when the microcode and chip layout were integrated for tapeout, and (2) providing another set of stimuli for finding errors in the microsimulator.

External interrupts have traditionally been the downfall of processor testing because of their asynchronous and unpredictable behavior. To alleviate this, an interrupt generator was added to the microsimulator, such that a test suite can contain directives requesting that an interrupt be generated

- at a specified point in the execution of a particular instruction
- at a random point
- at all possible points in the execution of a particular instruction

Although much of the standard architecture test suite applies to the microcoded functions, considerable resources were spent on instrumentation of the microcode and microsimulator to improve the test suites. This instrumentation performs coverage analysis, or feedback to the tester of situations that need further test coverage.

The coverage analysis tools allow one to check, across the duration of all test suites, that

- each microinstruction has executed at least once

- each conditional branch in the microcode has executed at least once in both directions

- each numerical comparison in the microcode has produced all three outcomes (less, equal, greater) at least once

- an interrupt was generated at the boundary of each microinstruction at least once

- each microinstruction in a microcode subroutine has executed at least once for all combinations of call history

- each conditional branch in a microcode subroutine has executed at least once in both directions for all combinations of call history

- each numerical comparison in a microcode subroutine has produced all three outcomes (less, equal, greater) at least once for all combinations of call history

- an interrupt was generated at the boundary of each microinstruction in each microcode subroutine at least once for all combinations of call history

The last four cases serve as a type of path-coverage analysis. For instance, if there is a subroutine in the microcode with the function of enqueuing a message on a queue and if this subroutine is called from more than one place in the microcode, the coverage analysis tools determine the test coverage of conditions in the subroutine for each context in which it is called.

CHIP DEBUG

Intel has always invested heavily in the development of proprietary CAD tools for VLSI development, and the sophistication of the tools was sufficient to raise hopes for what was called "A-step functionality," meaning that complex VLSI chips containing hundreds of thousands of transistors would work the first time. "Work the first time" did not merely mean being functionally correct; it also implied an expectation that the products would pass the tougher tests of correct behavior across their temperature and frequency specifications. These hopes had been realized with the first 80386 chips, and were realized again with the 80960. Within two days after production of the first A-step wafers, the 80960MC chips were successfully running architecture test suites. The A-step chips were eventually used in prototype systems and were shipped as operable samples to some customers.

The A-step chip, although highly functional, was not without bugs. By running architecture verification suites, and because the chip was functional enough to be used by software developers, about 40 bugs were eventually uncovered. About 80% were microcode problems and 20% logic and circuit problems, or about one bug per 10,000 transistors. Most of the bugs were sufficiently obscure that they did not affect software development. Some had a larger impact but were dealt with by "work-arounds" placed in the code generators of the compilers. A few were serious enough or impeded the debugging of other aspects of the chip to such an extent that an immediate action was needed (e.g., a short on one internal bus line in the floating-point unit rendered the floating-point unit untestable in the initial wafers).

The technique used as a short-term measure to eliminate serious bugs was the production of *partial* steppings when a bug could be fixed by altering only one mask layer of the chip. Since producing a new mask layer would consume too much time (but was nonetheless always done in parallel as a backup), the changes were made by identifying the physical change to the mask and then physically altering the mask. If the change required the removal of chrome from the mask, "laser zapping" was used. Laser zapping is the technique using a laser beam to burn away the necessary chrome from the mask.* If the change required the addition of chrome, an ion miller was used, which, instead of depositing chrome, produces the same effect by creating a prism in the glass of the mask. Three revisions of the A-step chip were produced in this manner.

Approximately six months after the receipt of the A-step chips, the second full tapeout of the processor chip was done to fix all of the A-step bugs, make a few small architectural changes, and fix some yield-limiting critical-speed paths. The silicon was available two months later. Although all of the changes were successful, the B-step chip experienced a high infant-mortality problem at high V_{CC} values because of the accidental creation of a zener diode within the chip. This was fixed by changing a single mask layer, resulting in the B-1 step. These chips were used extensively in internal prototype systems and used by customers.

Testing and usage of the B-step components turned up some additional bugs. The majority of these were associated with extremely unlikely circumstances and thus had little effect on the existing customer base. These were corrected in the C-step tapeout.

---

*Although the processor chip didn't require it, laser zapping on one of the other chips was used to correct bugs by actually modifying the chips themselves. When a bug could be eliminated or disabled by removing a metal line on the topmost metal layer of the chip, the lid of the chip package was removed and a laser beam was used to burn away the appropriate metal line or connection. The engineer who performed this process became so adept that he could also create connections where none existed by carefully burning two adjacent pieces of metal so that the splattered metal created a path between the two.

REFERENCES

1.   R. P. Weicker, "Dhrystone: A Synthetic Systems Programming Bench-mark," *Comm. of the ACM*, Vol. 27, No. 10, 1984, pp. 1013-1030.

2.   D. Budde *et al*, "A Methodology for VLSI Chip Design," *Lambda*, Vol. 2, No. 2, 1981, pp. 34-44.

CHAPTER 8

80960 Performance

In this chapter we present information about the performance of the first-generation 80960 microprocessors, along with information pertaining to the optimization of programs (e.g., useful optimizations for compiler backends and for time-critical assembly-language programs).

Unlike simpler, earlier microprocessors, the 80960s do not have straightforward, static instruction times. Some of the reasons are

- instruction pipelining

- memory-interface pipelining

- memory (bus) timing

- register and condition-code scoreboards

- instruction cache

- translation lookaside buffer

- multiple sets of local registers

- data dependencies of instructions (e.g., the data-dependent times of most floating-point instructions)

- state dependencies of instructions (e.g., the process-management instructions)

- other implementation-dependent interactions between successive instructions

Rather than attempt to fold all of these complexities into a timing analysis or formula for each instruction, the best-case times for the instructions will be presented, and the implications of the items in the above list will be discussed separately.

Unless otherwise noted, instruction performance is discussed in units of cycles. For a 20 MHz processor (external clock frequency of 40 MHz), the duration of one cycle is 50 ns. Also, unless otherwise noted, we assume that the pertinent instructions are resident in the instruction cache and that all memory references "hit" the TLB. The consequences of instruction-cache and TLB misses are discussed separately. Finally, unless otherwise noted, we assume a zero-wait-state memory configuration. This means, for instance, that when an address for a read is issued in cycle i, the data is returned in cycle $i+1$ for a one-word read and in cycles $i+1$ through $i+4$ for a four-word read. Again, the impact of adding wait states is discussed separately.

All the data herein applies not to the architecture, but to the 80960KA, 80960KB, and 80960MC initial implementations of the architecture. As discussed in Chapter 6, although the 80960 K-series processors are high-speed processors, their design contains many performance compromises (for reasons of schedule and space), and they do not exploit many of the opportunities purposely placed in the architecture (e.g., concurrent instruction execution). It is likely that the next implementation of the architecture will achieve speeds of three to four times greater.

LOW-LEVEL INSTRUCTIONS

The 80960s are designed so that the most frequently used instructions are executed in a single clock cycle (e.g., 50 ns for a 20 MHz processor); these instructions are listed in Table 8-1.

Table 8-1:   One-cycle instructions

mov	addi	addo
subi	subo	scanbyte
addc	subc	rotate
shlo	shro	shri
cmpi	cmpo	concmpi
concmpo	ornot	nand
and	notand	andnot
xor	or	nor
xnor	not	notor
chkbit		

Since the register array is single-ported (two registers cannot be read in the same cycle), there is one situation in which these instructions can take two cycles: when both operands are registers and the B (i.e., second operand) register is not the same register as that of the previous register destination. The ADDO instruction is used below to illustrate this situation.

```
addo    1,r4,r4        # 1 cycle
addo    r3,31,r10      # 1 cycle
addo    g0,r10,g0      # 1 cycle (B [r10] was previous destination])
addo    g9,g8,g8       # 2 cycles
```

Our studies have shown that, for these instructions, the most common cases are those in which (1) one operand is a literal and (2) one operand is the same as the destination of the previous instruction. Therefore, statistically, these instructions execute in $1+\varepsilon$ cycles, where ε is close to zero. The latter case, called the *bypass optimization*, is an important "peephole" optimization for compilers and assembly-language programs. The rule of thumb is - where both input operands of an instruction are registers and where the order of the operands is not significant, order the operands so that the B register is the same register as the previous register destination.

To illustrate the effectiveness of this tradeoff, in the Dhrystone[1] benchmark program 87% of the two-input instructions in Table 8-1 execute in one cycle; 13% of the instructions have two input registers where the B register is not the same as the most recent register destination. Therefore, the average time of these instructions in this program was 1.13 cycles.*

The other exception to the single-cycle timing of the instructions in Table 8-1 applies to the ADDC, SUBC, CONCMPO, and CONCMPI instructions. These instructions use the condition code as an input. If the previous instruction is one that sets the condition code, these instructions take an extra cycle; otherwise they are executed in one cycle.

The speed of the other low-level instructions is shown in Table 8-2. A few of the instructions have data-dependent speeds (e.g., SCANBIT and SPANBIT, which search for the most significant 0 or 1 in a register).

The multiplication instructions have the widest range of possible times. In addition to performing the multiplication at two bits per cycle, the processor performs a lookahead at the next three and seven bits of the multiplier (the B operand). If they are all 0 or 1, it can process them in a single cycle. As a few examples, 0×0 takes 9 cycles, 100×10 takes 12 cycles, and 5555555516 multiplied by itself takes 21 cycles. Because the lookahead applies to the B operand, one should reorder the A and B operands, when the opportunity presents itself, to have the B operand be the one with the most consecutive 0's and 1's. For instance, if one of the operands is a constant, it is usually wise to make it be the B operand.

Note that the SHLI instruction consumes two cycles, while SHLO and SHRI consume one cycle. The reason for the difference is that SHLI can produce an overflow fault.

In the 80960s, there is no special hardware for bit-field extraction, and the EXTRACT and MODIFY instructions are implemented in microcode, leading to their relatively slow speeds. Since EXTRACT handles the most general case (where the offset and length of the bit field are not necessarily static), a sequence of other instructions will outperform it in many cases. For instance, if the offset and length are fixed (as the values F and L), the following instruction sequence performs an extract in two cycles.

```
shlo        32-L-F,reg,reg
shro        32-L,reg,reg
```

---

*We exclude the MOV instruction, since it always executes in a single cycle. If we count the MOV instructions, the ratio rises to 90%, or 1.1 cycles per instruction.

Table 8-2: Multiple-cycle instructions

Instruction	Cycles
movl	2
movt	3
movq	4
mulo,muli	9-21
divo,divi	41
remo,remi,modi	43
emul	16-27
ediv	49
setbit,clrbit	2
notbit	2
alterbit	2
scanbit,spanbit	7-14
extract	7
modify	8
modac	12
dmovt	7
daddc,dsubc	8
shli	2
shrdi	3
cmpinco,cmpinci	2
cmpdeco,cmpdeci	2
syncf	3

Where the values are not static, the following sequence is equivalent to the EXTRACT instruction:

```
shro    g0,g2,g2    # sequence equivalent to extract  g0,g1,g2
shlo    g1,1,g3
subo    1,g3,g3
and     g2,g3,g2
```

The alternative sequence takes four cycles instead of seven. However, because of the way the instruction cache works, this sequence will actually take five cycles (the reason will be discussed later). Furthermore, the sequence uses an additional register (g3) and consumes four words of space instead of one. Therefore, the

EXTRACT instruction should generally be used unless both the bit-field offset and length are constants.

As discussed earlier, instructions that have two source operands can take an additional cycle if both source operands are registers and the B register is not the same register as that used most recently as a register destination.

One other source of additional time for the instructions in this section is the cost of overflows when integer overflow is masked. The 80960's ALU does not use the overflow mask flag. Instead, when an overflow occurs, the processor traps to microcode, which determines whether overflow is masked. As a result, 13 additional cycles are expended whenever an integer instruction overflows and overflow is masked.

BRANCHING

To understand the timing of branch instructions, it is helpful to understand how they are implemented. Branch instructions are handled by the instruction decoder, possibly in parallel with the execution of the prior instruction in the execution unit (if the branch doesn't depend on the condition code, or if it does depend on the condition code but the previous instruction is not one that alters the condition code). To keep track of the latter condition, the processor maintains a *condition-code scoreboard*, an internal flag signifying whether the current instruction will alter the condition code. The execution times of the branch instructions, and a few other related instructions, are shown in Table 8-3.

The unconditional branch (b) takes from zero to two cycles, depending on the previous instruction. If the previous instruction is not a branch and consumes one cycle, the branch instruction will consume one additional cycle. If the previous instruction is not a branch and consumes two or more cycles, the branch will be completely overlapped and take no additional time.

The conditional branch timings depend on whether the branch is taken and whether the previous instruction alters the condition code. For instance, the following sequence consumes a total of five cycles if the branch is taken, and four if it is not.

```
mov     0,r4
cmpo    0,r5
bne     xyz
```

Table 8-3:   Branch instruction times

Instruction	Cycles	Condition
b	0-2	
be,bne,...	0-2	branch not taken
	0-3	branch taken
cmpobe,...	3	branch not taken
	4	branch taken
bbs,bbc	3	branch not taken
	4	branch taken
bx	4	addressing mode is 32-bit displacement
	4	addressing mode is (reg)
bal	3	
balx	5	addressing mode is 32-bit displacement
balx	5	addressing mode is (reg)
faulte,...	2	no fault occurs
teste,...	2-3	

Since the CMPO instruction is independent of the MOV instruction, one can reduce the time of the sequence by one cycle by reordering the instructions in the following way.

```
cmpo    0,r5
mov     0,r4
bne     xyz
```

If one can move additional instructions between the compare and the branch, or move a multiple-cycle between the two, the effective cost of the branch can be reduced to zero. This is an important optimization for compiler writers and assembly-language programmers: Wherever possible, try to reorder code so that there are one or more cycles of execution between an instruction that sets the condition code and a subsequent instruction that tests the condition code.

Seemingly in contradiction with the above suggestion are the compare-and-branch instructions, which provide an inseparable compare and branch. These instructions have the same timing as a compare followed by a conditional branch.* They fit into the scheme of things in the following way: If there are one or more

---

*And they consume an additional cycle if two registers are specified and the B register is not the same as the previous register destination (as explained in the previous section).

instructions that can be moved between a compare and conditional branch, separate compare and branch instructions should be used. If there are not, the compare-and-branch instructions should be used, in spite of the fact that they appear to take the same time as a compare immediately followed by a branch. The reason is that these instructions are a space optimization (taking one word instead of two), and therefore will improve the instruction-cache hit ratio.*

A similar situation applies to the BBS and BBC instructions. Whether one should use these or an equivalent sequence of a CHKBIT instruction and a conditional branch depends on whether possibilities for reordering instructions can be found; if an instruction cannot be moved between the CHKBIT and branch, use BBS or BBC instead.

The BX and BALX instructions use the same addressing modes as the load/store instructions. Table 8-3 shows the two most common cases: where the branch address is a 32-bit value following the instruction, and where the branch address is in a register.

Although they are not branching instructions, the FAULT and TEST instructions are shown in the table because they depend on the condition code. The FAULT instructions always consume two cycles, and the TEST instructions consume three cycles if the previous instruction sets the condition code, and two cycles if not. The following two sequences compare the use of the TEST instructions to the equivalent code had not the TEST instructions existed. The longer sequence takes four or five cycles (depending on the result of the comparison), whereas the comparison and test take four cycles (and half of the space).

```
          cmpo    7,r4              cmpo    7,r4
          mov     1,r5              teste   r5
          be      xyz
          mov     0,r5
xyz:
```

LOAD/STORE INSTRUCTIONS

Load/store timing, which one might think of as straightforward, is extremely complex in the 80960s. Rather than give instruction speeds, which is virtually impossible to do, we will show how the processor performs these instructions, give examples of timings, and discuss a set of software optimizations.

---

*One exception to this suggestion is the rare occurrence where the branch displacement is too large to be encoded in these instructions.

The timing of loads and stores is complex because it depends on the following factors, many of which are interdependent:

1. the time to calculate the effective address (i.e., the addressing mode used)

2. whether the TLB is hit

3. for loads, when the first subsequent instruction arrives that uses the loaded data (i.e., the use of the register scoreboard)

4. for stores, the status of the write buffers in the processor

5. the number of wait states in the bus transaction

6. the alignment of the memory operand

7. the state of the bus (i.e., whether the bus is busy with a prior load/store, an instruction cache prefetch, or an I/O transfer)

8. the nature of the n subsequent instructions (i.e., whether they access the register array)

To begin to sort out the effect of these factors, the following list shows the timing of an LD instruction, assuming zero wait states, a TLB hit, alignment of the operand on a word boundary, a nonbusy bus, the next few instructions not depending on the result of the load, and an addressing mode having a zero effective-address calculation time.

Cycle	Processor activity	Bus activity
1	See LD inst and translate address	
2	Execute inst $i+1$	T_a (address) cycle
3	Execute inst $i+2$	T_d (data) cycle
4	Execute inst $i+3$	Transfer load data to register

This shows that the best-case time of a load instruction is one cycle. However, since the processor does not have a separate port into the register array for load/store operations, the transfer of the loaded data in cycle 4 will delay the execution of an instruction in that cycle, unless the instruction being executed does not require use of the register array and a certain control bus during that cycle.

Examples of cases where there is no delay are a branch instruction, a multiply instruction (if started in cycle 2), and other load instructions (depending on their addressing modes). If the LD instruction used an addressing mode requiring an effective-address calculation of one or more cycles, these cycles would appear at the beginning. Thus the best-case effective time consumed by a LD instruction is given by

$$load\_time = efa\_time + 1 + transfer\_time$$

where efa_time could be zero (but is typically 1 or 2), where the additional cycle is for address translation*, and where transfer_time is usually 1 (but occasionally 0). If the instruction following the load is dependent on the loaded data, the effective time consumed by the LD instruction is

$$load\_time = efa\_time + 1 + transfer\_time + bus\_time$$

where transfer_time is always 1 and bus_time is 2 for 0 wait states. Table 8-4 shows the effective-address calculation times.

Table 8-4:   Effective-address calculation times

Cycles	Addressing mode
0	12-bit offset (MEMA format)
1	displacement (32 bits)
1	(reg)
2	12_bit_offset + (reg) (MEMA format)
2	displacement + (reg)
2	displacement + (reg) $\times 2^{scale}$
3	(reg) + (reg) $\times 2^{scale}$
3	displacement + (reg) + (reg) $\times 2^{scale}$
4	displacement + IP + 8

The timing of the byte and halfword load instructions is the same as that of the LD instruction. The timing of the LDL, LDT, and LDQ instructions is similar, except that their transfer times are 2, 3, and 4, respectively (although some or all of the transfer time is occasionally hidden), and their bus times are 3, 4, and 5. The following examples show the effective times for a variety of situations.

---

*The address-translation cycle is always present, even if address translation is disabled, and even in the 80960KA and 80960KB.

x1:	ld	64,r4	# 4 cycles
	addi	1,r4,r4	# 1 cycle
	...		
x2:	ld	(r4),r5	# 5 cycles
	addi	1,r5,r5	# 1 cycle
	...		
x3:	ld	(r4),r5	# 5 cycles
	addi	r4,r5,r5	# 1 cycle (note: not 2 cycles)
	...		
x4:	ldq	(r4),r8	# 8 cycles
	addi	1,r8,r8	# 1 cycle
	...		
x5:	ld	64,r4	# 2 cycles net
	addi	1,r8,r8	# 1 cycle
	mov	0,r9	# 1 cycle
	cmpi	0,r7	# 1 cycle
	addi	1,r4,r4	# 1 cycle
	...		
x6:	ld	0,g2	# 2 cycles net
	ld	20,g3	# 3 cycles net
	ld	(g1),g4	# 3 cycles net
	ld	(r15),g5	# 3 cycles net
	ld	(g2),g6	# 4 cycles net
	extract	6,g3,g3	# 7 cycles

Examples x1 and x2 are straightforward. In both cases the next instruction needs to be delayed until the load completes. Example x3 is similar, except that it is followed by an instruction that could consume two cycles. This example shows that the bypass optimization applies to loads. Example x4 shows the timing of a four-word LDQ instruction where the next instruction is dependent on one of the registers loaded.

Example x5 is important because it illustrates use of the register-score-board optimization. When a load instruction is executed, the processor marks the destination register(s) as busy and continues instruction execution (in parallel with the bus operation). If an instruction's register operands are not busy, the instruction is allowed to execute; otherwise the instruction is delayed until the registers' status changes to nonbusy. When data from the bus is moved into a register as a result of a load, the processor marks the register as nonbusy. In example x5, we've found three cycles of work to move between the load and the first instruction that requires the loaded data. As a result, the effective time taken by the load instruction is two cycles instead of the four taken in example x1 (or looking at it another way, we have executed two cycles of intervening instructions "for free").* This type of optimization can have a significant impact on performance; where possible, try to find instructions that can be moved between a load instruction and the first instruction that uses the destination of the load.

The last example shows a series of independent load instructions. Because of the protocol of the multiplexed external bus (the bus cycle following the last data cycle for a read operation is an idle cycle, which is needed to turn the bus drivers around), the timing of a series of load instructions is constrained by the bus timing.

Before leaving the subject of the register scoreboard, two quirks in the way it functions for LDL and LDT instructions should be explained. When a LDT instruction is executed, an additional register is marked as busy as if the LDT were a LDQ. For instance, in the following sequence,

```
ldt     xyz,r8
addi    1,r11,r11
```

the ADDI instruction will be delayed because the processor marks r8-r11 as busy. For the LDL instruction, the scoreboarding is more complicated. If the destination register operand (the first of the pair) is a multiple of 4 (e.g., g0, g4), the scoreboard is marked as if the LDL were a LDT or LDQ (e.g., if the destination is g0, g0-g3 are marked as busy). If the destination register is not a multiple of 4 (e.g., g2, g6), the following applies: If the LDL destination is g2 or r2, then g0-g7 or r0-r7 are marked as busy. If the destination is g10 or r10, then g8-g15 or r8-r15 are marked as busy. However, if the destination is g6, r6, g14, or r14, all of g0-g15 or r0-r15 are marked as busy.**

We've noted several times that one or more of the net execution cycles of load instructions occurs when the storing of the loaded data into the register array creates interference with some subsequent instruction being executed. One can optimize a program to minimize this interference, although the optimization requires an exact understanding of the memory timing in one's particular system, and it is a probabilistic optimization in that it won't work all of the time because of factors that are difficult to predict (e.g., instruction cache prefetches, external cache misses).

The "trick" is to have an instruction that will not create interference executing in the exact cycle(s) that the data from a prior load instruction is being moved into the register array. The primary instructions that serve this purpose are branches, floating-point instructions, and the long-executing integer instructions (e.g., multiply

---

*The load has a net cost of two cycles instead of one because the movement of the data from the bus conflicts with one of the subsequent instructions. A way to avoid this and optimize the load time by one more cycle will be shown shortly.

**This doesn't make much sense. It is an outcome of making some architecture changes rather late in the implementation of the chips.

and divide). For all but the branches, one must also understand exactly when the instruction will not result in interference. For the integer instructions and 32-bit floating-point instructions, the first three cycles and the last cycle will create interference. For 64-bit floating-point instructions, the first five cycles and the last two cycles will interfere.

```
x1:   ldq     64,r4        # just 1 cycle net
      mulo    30,30,g0     # 11 cycles
      ...
x2:   ldq     (r4),r4      # just 2 cycles net
      mulo    30,30,g0     # 11 cycles
      ...
x3:   ld      (g7),g2      # this example executes in 8 cycles.
      b       x3a          # the branch is overlapped with the load, but the
      ...
x3a:  setbit  r14,r15      # storing of the load result into g2
      addo    1,g2,g2      # interferes with the setbit instruction
      ...
x4:   ld      (g7),g2      # this example, the same four instructions,
      setbit  r14,r15      # executes in 7 cycles because of the
      b       x4a          # placement of the branch
      ...
      addo    1,g2,g2
```

The timing of store instructions is simpler because they do not mark the register scoreboard, and because they access the register array at the beginning, rather than at some hard-to-predict later point in time. Using the same assumptions as for the LD instruction, the timing of a ST instruction is shown below.

Cycle	Processor activity	Bus activity
1	See ST inst and translate address	
2	Read store data from register	T_a (address) cycle
3	Execute inst $i+1$	T_d (data) cycle

Thus the time consumed by a ST instruction is given by

store_time = efa_time + 2

where efa_time is the same as that shown for load instructions (e.g., possibly 0, but typically 1 or 2). For the STL, STT, and STQ instructions, the time increases by one cycle each.

In terms of the bus activity, a zero wait-state store consists of a T_a cycle and a T_d cycle. However, although not a requirement of the bus protocol, the processor refuses to use the bus cycle following the T_d cycle of a write transaction. This means that the fastest rate at which consecutive stores can be done is three cycles. However, store instructions can be executed at a faster rate; up to three consecutive store instructions can be executed in two cycles each. The processor does this by implementing three write buffers, which can hold the addresses and data for three store operations. Each buffer can hold four data words so that STQ instructions can be buffered. This buffering decouples, to some degree, a program's execution rate through store instructions from the bus bandwidth. For instance, the following sequence of instructions executes in seven cycles.

```
st      64,(g5)
st      80,(r4)
st      100,(r5)
mov     r15,g0
```

This example may appear contrived, since it uses the one addressing mode that has a zero effective-address calculation time and this addressing mode is not frequently used. However, the example also assumes zero wait states, which is frequently not the case (e.g., where one has a zero wait-state external cache, but the cache is a write-through cache). In more likely cases, where the stores have a one- or two-cycle effective address time and where consecutive writes cannot be done to memory as quickly as one every three cycles, the write buffers provide a significant benefit.

Because of the register scoreboard for loads and the write buffers for stores, and because of other attributes, such as the large number of registers and the instruction cache, the 80960 processors are much less sensitive to external memory speeds than most other processors. A basic rule-of-thumb is that each incremental wait state in an 80960 design costs one about a 7% loss in performance, contrasted with a loss of 25% or more per wait state in other microprocessors. This means that an external cache is less important in an 80960 design than in other microprocessor-based designs, and that one can build a high-performance, cost-effective, system using just DRAMs.

One of the assumptions of the previous discussions is that the operands in memory are aligned. Since the 80960MC supports unaligned operands, it is worthwhile to examine the cost of unaligned accesses. In the following sequence, where the result of the load is used by the next instruction, the cost of the LD instruction is five cycles.

```
ld        (g0),g1
mov       g1,g2
```

If the address in g0 is not on a word boundary but does not cross a 16-byte boundary, the cost is only one additional cycle (the processor performs a two-word read operation on the bus). If the address is not on a word boundary and the word desired spans a 16-byte physical address boundary, the cost is four additional cycles (i.e., nine instead of five). Three of these additional cycles are the result of the bus protocol; the processor must issue two independent read operations. The fourth additional cycle is an additional address-translation cycle; when an access spans a 16-byte boundary, the processor assumes it could also cross a page boundary and as a result performs an address-translation cycle for each part of the word.

For the ST instruction, the cost is one cycle more than for the LD instruction. If the address is not a word boundary, the ST instruction takes two cycles longer, unless a 16-byte boundary is also crossed, in which case the ST instruction takes five cycles longer.

Another working assumption has been that all memory accesses hit the TLB (80960MC only). The TLB contains 44 entries; 32 of these are available for holding translation information for the most recently accessed pages. Given a page size of 4k bytes, the TLB maps 128k bytes of the address space. Previous studies[2] indicate that a TLB with these characteristics should achieve a hit ratio of 99.8%.

Because of chip-size limitations, the 80960MC traps to microcode on TLB misses and thus handles misses relatively slowly. If the region is mapped by a one-level page table, a miss consumes 17 additional cycles (assuming zero wait states). If it is mapped by two levels of page tables, a miss consumes 31 cycles.

These numbers assume that the accessed and altered flags in the page-table or directory entry are already set. If the miss requires that the processor set one or both of these flags, another 20 cycles must be added. The reason for this large number is that the architecture requires that these flags be set in memory using a read-modify-write operation to ensure correct behavior in multiple-processor systems. However, since in the typical case these flags are already correct, the processor first reads the entry with a normal read operation and then rereads it, if necessary, if a read-modify-write cycle is required.

Taking all of this into account, TLB misses add an average of about 1.5-2% to the cost of memory accesses in the 80960MC.

Table 8-5 shows the speeds of additional instructions related to load/store instructions. For the LDA instruction, efa_time is that listed in Table 8-4, implying that the timing of LDA is from one to five cycles. An instruction such as

```
lda       10(r8),r8
```

takes three cycles. This means that one should not use LDA to add small constants to registers; one should instead use the one-cycle ADDO instruction.

Table 8-5:   Related instruction times

Instruction	Cycles
ldphy	17
inspacc	29
atadd	18
atmod	21
lda	1 + efa_time

INSTRUCTION-CACHE CONSIDERATIONS

A major contributor to the performance of the 80960 K-series processors is an on-chip instruction cache. Most readers will be interested only in the hit ratio of the instruction cache and how that affects the average instruction time. However, for those interested in achieving the optimal performance for some inner algorithm, and to warn about some possible worst-case performance effects, we will discuss the organization and operation of the cache.

The cache is organized as a 512-byte, direct-mapped cache. Direct mapped means that an instruction at any arbitrary address in memory can occupy only one position in the cache. This is shown in Figure 8-1. The cache consists of 32 blocks (or lines), each of which contains 16 bytes (4 instructions). An instruction residing in a 16-byte block of memory whose address is i can only occupy line $i/16$ *mod* 32 in the cache, and thus blocks whose addresses differ by a multiple of 512 cannot coexist in the cache.

Block   Mapped addresses

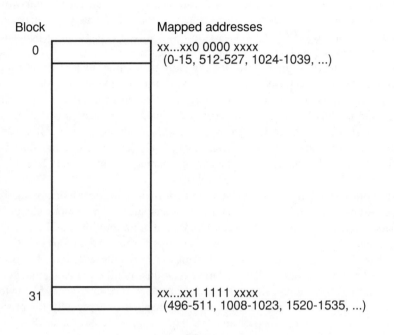

Figure 8-1: Instruction cache organization

To make optimum use of the memory bus with its multiword transfer capability, the lines are 16 bytes in size, and a miss in the cache always fetches 16 bytes from memory (rounding the address down to the nearest 16-byte boundary and using the four-word read operation of the bus). In addition to fetching on demand, the cache logic performs two types of anticipatory prefetches, which have been shown to be of significant benefit to performance because of the sequential nature of program flow.[3] When there is a cache miss and, as a result, the processor needs to perform a 16-byte fetch from some address i, the processor immediately initiates a second fetch (a prefetch) of the block at address $i+16$. Also, whenever a branch is taken to some instruction that happens to reside in the cache, the processor immediately issues a prefetch of the next 16-byte block.

The most obvious attribute of the instruction cache is its hit ratio, or the probability for each instruction executed that the instruction is already in the cache. For large programs, the hit ratio is typically in the range of 88-90%, a ratio that matches others' results.[4,5] One study[5] predicts a hit ratio of 83% for a 512-byte direct-mapped instruction cache, but it assumes that no prefetch is used.

With a hit ratio of approximately 90%, the speed with which a miss is handled is critical, since this occurs on an average of 1 of every 10 instructions executed.

After the processor issues a four-word read, it does not wait for all four words to arrive; as soon as the needed instruction arrives, it is immediately decoded and executed.* The speed with which a cache miss is resolved is therefore dependent on a number of factors, such as the availability of the memory bus (e.g., whether there is a prior cache fetch or prefetch, or load or store instruction, underway), the number of wait states, and the position of the needed instruction in the block of four words. For a zero wait-state configuration, where the needed instruction is the first of a block and the bus is not busy, the cost of an instruction cache miss is six cycles. The typical cost of a miss (for zero wait states) is about eight cycles. If we assume that for an infinitely large instruction cache the average instruction time is two cycles, the actual cache results in an average instruction time of $0.9(2) + 0.1(2+8)$, or an average instruction time of 2.8 cycles.

For those concerned enough about the performance of inner loops or algorithms to be willing to worry about how their program is laid out in memory, the principal situation to avoid is jumping between blocks that are a multiple of 512 bytes apart. The direct mapping of the cache leads to ideal performance for loops or algorithms less than 512 bytes in size (independent of the beginning address). Performance drops, of course, for loops that are larger than this, but the prefetch strategy holds the drop to a minimum. Although control transfers of a multiple of 512 bytes are rarely seen in the case of branches, they can happen unexpectedly for procedure calls. If the small inner loop of an algorithm contains a call to a small procedure that by chance happens to be mapped 512 bytes away, the instruction cache will be seen at its worst.

Given the above warning, one might ask why the cache is direct mapped as opposed to being fully or n-way associative. There are three reasons. The first is silicon area; direct mapping is the cheapest to implement. The second is that for a range of programs simulated, we found that a direct-mapped cache often outperformed a two-way set-associative cache, and in the worst case was only marginally slower (a result confirmed elsewhere[5]). The third is that a set-associative cache creates similar performance anomalies.

To illustrate the latter point, Figure 8-2 shows a cache of the same size (512 bytes), organized as two-way set associative. There are still 32 lines in the cache, but an instruction can occupy one of two slots in the cache. A replacement algorithm is needed to determine, on a fetch or prefetch, which of the two slots should be used.

---

*Because of this, when one is interested in squeezing every microsecond out of a program, one should align branch targets on 16-byte boundaries whenever possible. For the same reason, a good general rule is to align procedures on 16-byte boundaries.

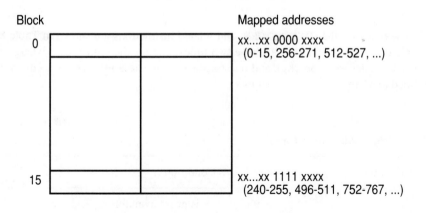

Figure 8-2: Hypothetical two-way organization

Consider a loop slightly larger than 512 bytes, where we execute a few instructions at address 0, jump to a few instructions at address 256, jump to a few instructions at address 512, and jump back to the beginning (i.e., there are three blocks of interest). In the direct-mapped cache, there is one cache miss per iteration of the loop. In the set-associative cache, there are three misses per iteration (one for every jump) if the replacement algorithm is LRU or FIFO.* If the replacement algorithm is random (we use a random bit to pick between the left and right slot), there is an average of two misses per iteration.

Finally, given that the instruction cache is not infinitely large, it is important to realize that it can influence instruction timings in three ways. The first, the obvious one, is the effect of the hit ratio and cache misses. The second is that prefetches can use bus cycles and thus can interfere with the execution of load/store instructions. The third is that when the cache logic is in the process of deciding whether to prefetch, it needs to determine if the block in question is already in the cache. This situation arises when the processor begins accessing instructions in a new cache line, either as the result of a branch or by "falling through" from the previous line. When this occurs, there will occasionally be an extra cycle of delay, depending on the timing of the current instruction.

---

*LRU, or least recently used, means that we fetch the new block into the slot (of the two possible slots) that was least recently referenced. FIFO means that we use the slot that was previously filled first.

CALL/RETURN

The times for the CALL, RETURN, and related instructions are shown in Table 8-6. Note that CALL and RETURN have two times: one for the typical case where the needed register set is on-chip, and one for the case where a register set needs to be flushed to, or retrieved from, memory.

Table 8-6: Call/return times

Instruction	Cycles	Condition
call	9	register set available
call	41	register set needs to be flushed
ret	8	caller register set on chip
ret	43	caller register set needs to be fetched
callx	12	register set available, addressing mode is (reg)
calls	47	register set available, switch from user to supervisor mode
ret	22	return from supervisor mode to user, caller register set on chip
flushreg	9	nothing to flush
flushreg	41	one register set to flush
flushreg	105	three register sets to flush

As with many other aspects of the architecture, there is ample opportunity to improve the speed of the call/return operations in future implementations. For instance, a trial design has proved the feasibility of doing a call instruction in two cycles and a return in three (and at a much higher clock frequency).

Because the call/return functions are closely tied to the usage and allocation of registers, it is worthwhile to examine how Intel's compilers view the registers. Table 8-7 shows how the call/return model implemented by the compilers interacts with the global registers.

Table 8-7:   Global register conventions

Register	Usage	Can be destroyed by callee
g0	parameter 0 / return 0	yes
g1	parameter 1 / return 1	yes
g2	parameter 2 / return 2	yes
g3	parameter 3 / return 3	yes
g4	parameter 4	yes
g5	parameter 5	yes
g6	parameter 6	yes
g7	parameter 7	yes
g8	parameter 8 / temp	yes only if parameter
g9	parameter 9 / temp	yes only if parameter
g10	parameter 10 / temp	yes only if parameter
g11	parameter 11 / temp	yes only if parameter
g12	temp	no
g13	structure return pointer / temp	yes
g14	leaf return pointer / argument pointer	no
g15	fp (frame pointer)	

The first important point is that parameters are normally passed in the global registers, rather than pushed on the stack in memory. Up to 12 registers may be used for parameters, which covers virtually all cases. In the rare case where there are more than 12 words of parameters, the caller (calling procedure) must place the remaining ones in memory (on the stack or otherwise) and pass the beginning address in register g14. If the procedure returns a value, the return value is placed in register g0, or g0-g4 for a multiword value. If the procedure returns a structure, the structure's address is returned in register g13.

To discuss the remaining conventions, it is necessary to distinguish between *leaf* and *nonleaf* procedures. A leaf procedure is one that calls no procedures; a nonleaf procedure calls other procedures. A nonleaf procedure typically* copies the first n global registers into its local registers, where n is the lower of (1) the number of words of input parameters and (2) the maximum number of parameter words it passes to a procedure it calls. However, if it passes 8-11 parameter words to another procedure, it must save and restore the corresponding global registers for its caller.

---

*"Typically" means that a compiler that performed extensive data-flow analysis could do better than this.

The nonleaf procedure uses the remaining local registers for local variables.* If the number of local variable words exceeds 13 (the number of usable local registers) minus the number of saved global registers, the remaining local variables are allocated on the stack in memory.

A leaf procedure leaves its input parameters in the global registers and can allocate local variables and temporaries in registers g0-g13 and r3-r15.

Note that the global registers are divided into two classes: those which a nonleaf procedure must assume are destroyed by a called procedure (and thus can be used at will by a called procedure) and those which cannot be destroyed by a called procedure (and thus must be saved by the called procedure if altered by it). Other compilers for other architectures typically do not provide these two usage classes; they adopt one convention or the other. However, since the 80960 architecture provides an ample number of registers, providing both classes of usage conventions minimizes the saving and restoring of registers.

As an illustration, consider a nonleaf procedure that receives four words of input parameters and passes six words of parameters to a called procedure. This procedure, upon entry, would move g0-g3 into local registers r3-r6 and use local registers r7-r15 for local variables. The procedure's code can then use global registers g6, g7, and g13 as temporaries. It can also use g4 and g5 as temporaries, providing that their values are not needed across the call to the called procedure. After the input values are no longer needed, the procedure can also use g0-g3 as temporaries. Finally, if it needs additional registers, it can also use g8-g12 and g14, but it needs to save them first and restore them before returning.

There is another optimization provided by Intel's compilers, assembler, and linker for leaf procedures, where the linker can change certain calls to separately compiled procedures to BAL instructions. For instance, a leaf procedure with only a few input parameters and a few local variables may not need its own set of local registers. In this case, if the compiler can map the input parameters, local variables, and any needed temporaries into global registers g0-g7 and g13, the compiler can mark the procedure (in its object-module file) as eligible for this optimization. When this procedure is linked to calling procedures, the linker will change any CALL instructions to BAL instructions if the CALL instructions have also been marked as eligible for the optimization.

---

*Only if the program does not take the address of one of these local variables. Local variables of this type must be allocated on the stack in memory.

FLOATING POINT

Expressing the speed of the floating-point instructions of the 80960KB and 80960MC is another difficult task because the times depend on a large number of factors such as the specific operand values, the length (precision) of the operands and result, whether the operand values have the same exponent, whether the result needs to be normalized, whether the result needs to be rounded, whether special input values (e.g., NaN, ∞) are used, and so on. For the sake of simplicity, a set of times are given in Table 8-8 using the following assumptions

- Unless otherwise noted, all input operands have the arbitrary value $\pi/3$. This value was precomputed to the full precision of the input operand (e.g., for 80-bit extended-real operands, $\pi/3$ was computed to 64 bits of precision).

- In the arithmetic-controls register, all the masks and sticky bits are set prior to executing the instruction.

- For 32- and 64-bit operations, the operands are in the 32-bit global or local registers (pairs of registers for 64-bit operands). For the 80-bit operations, the operands are in the floating-point registers.

- No mixed-precision operations are shown.

Also, since these times, and the times in the remaining sections of this chapter, tend to be large, we will depart from the convention of giving them in cycles and express them instead in microseconds (assuming a 20 MHz processor).

Table 8-8:   Floating-point times (microseconds, 20 MHz)

Instruction	32 bits	64 bits	80 bits	Notes
movr			0.3	source is 32 bits
movr	0.3			source is 80 bits
movrl			0.4	source is 64 bits
movrl		0.4		source is 80 bits
movre			0.4	
movre			0.5	src. is 80 bits in global/local regs
movre			0.5	dest. is 80 bits in global/local regs
add	0.5	0.7	0.5	
sub	0.5	0.6	0.5	
mul	1.0	1.6	1.8	
mul	0.6	0.8	0.6	minimum time (B operand is zero)
div	1.9	3.9	3.8	
rem	3.7	3.8	3.7	
scale	1.6	1.7	1.6	A operand is +2
round	4.3	4.3	4.1	
sqrt	4.9	5.0	4.9	
sin	18.8	20.9	20.9	
cos	18.9	21.0	21.0	
tan	13.6	15.6	15.4	
atan	7.8	12.9	12.6	
exp	13.3	13.6	13.3	A operand is $\pi/3 - 1$
logbn	1.4	1.5	1.4	
log	20.8	21.4	21.4	
logep	20.5	21.0	21.0	A operand is $\pi/3 - 1$
class	1.2	1.2	1.2	
cpysre			0.8	
cpyrsre			0.9	
cmp	0.6	0.7	0.7	
cmpo	0.6	0.7	0.6	
cvtri	2.2		2.2	
cvtril	2.1		2.1	
cvtzri	2.0		2.0	
cvtzril	1.9		1.9	
cvtir	1.9		1.9	A operand is +2
cvtilr	1.8		1.8	A operand is +2

STRING OPERATIONS

The string instructions of the 80960MC copy, fill, and compare strings of bytes when the length of the string is known in advance (as opposed to string functions in the C language, which are delimited by a terminator byte and are best implemented by using the SCANBYTE instruction). Since copying of nonoverlapping data is the most frequently used operation, and since the 80960MC's MOVQSTR instruction is not always the fastest alternative, the speeds of different alternatives for different situations are compared in Table 8-9.

The three alternatives are the MOVQSTR instruction, a software loop, and an "unfolded" software loop. The last consists of one or more sets of load and store instructions, possibly combined with intermixed addition instructions to increment the addresses, and is viable only when the size of the string is small and known at compilation or program-writing time. The second alternative, the loop, consists of the following code and assumes the string is a multiple of 16 bytes in size:

```
        mov       srcreg,r4
        mov       destreg,r5
        shro      4,lengthreg,r6
top:    ldq       (r4),r8
        addo      16,r4,r4
        cmpdeco   1,r6,r6
        stq       r8,(r5)
        addo      16,r5,r5
        bl        top
```

Two string lengths are shown in the table - 16 and 64 bytes - as well as each incremental 64 bytes beyond the initial 64 bytes. For each string length there are three cases of alignment of the source and destination: both begin on 16-byte boundaries, both begin on word boundaries (but not 16-byte boundaries), and both begin on nonword boundaries.

Table 8-9: String copy times (microseconds, 20 MHz)

Size	Alignment	MOVQSTR	Software loop	Unfolded sequence
16 bytes	quadword	1.6	1.0	0.7
	word	1.8	1.1	0.9
	byte	2.1	1.4	1.1
64 bytes	quadword	4.7	3.8	3.4
	word	5.7	4.4	4.3
	byte	7.7	5.3	5.0
each 64 bytes	quadword	3.5	3.3	
	word	3.5	3.8	
	byte	4.3	5.2	

Although MOVQSTR is not the fastest alternative in most of the cases, it is probably the best choice in most circumstances. For instance, the software loop consumes more code space, uses seven working registers, and makes the simplifying assumption that the string is a multiple of 16 bytes.

For nonoverlapping strings (the assumption made in the table), the MOVSTR instruction is 0.5 μsec. (10 cycles) slower than MOVQSTR.

Table 8-10 shows the speed (at 20 MHz) of overlapping string moves, as well as the FILL and CMPSTR instructions.

Table 8-10: Other string times (microseconds, 20 MHz)

Instruction	Size	Alignment	Time	Notes
movstr	64 bytes	any	45.1	dest. overlaps, any boundary
movstr	each byte	any	0.7	destination overlaps
fill	64 bytes	quadword	2.9	
fill	64 bytes	word	2.9	
fill	128 bytes	word	4.7	
cmpstr	64 bytes	quadword	12.8	strings equal
cmpstr	64 bytes	quadword	2.3	strings unequal in first byte
cmpstr	64 bytes	byte	15.4	strings equal
cmpstr	64 bytes	byte	2.3	strings unequal in first byte

PROCESS MANAGEMENT

Table 8-11 shows the speed of instructions and other operations associated with the process management functions in the 80960MC. Unless otherwise noted, the functions that involve suspension of a process assume that only one set of local registers is in use; about 3.6 μsec. must be added if all four sets are in use. The times for instructions that refer to specific segments (e.g., ports, semaphores, messages) assume that the segment-table entries for these segments are already in the TLB. Since this is relatively unlikely, the times should be increased by 1.2 to 3.8 μsec. The variation depends on whether one or two segments are referenced and whether their accessed and altered flags need to be set. Also, in many of the numbers there is a small degree of variability depending on the state (e.g., whether a message is sent to an empty or nonempty port).

Table 8-11: Process-management function times (microseconds, 20 MHz)

Instruction/Operation	Time	Notes
send to FIFO port	5.0	
send to priority port	5.7	
sendserv	15.4	includes process suspension
schedprcs	6.7	assumes process is nonpreempting
receive from FIFO port	5.2	port has a message
receive from priority port	6.1	port has a message
receive blocks	16.3	includes process suspension
condrec	3.7	port has no messages
condrec	4.9	port has a message
signal	2.0	
wait	2.3	semaphore has a signal
wait blocks	15.7	includes process suspension
condwait	2.6	semaphore has no signals
condwait	2.3	semaphore has a signal
ldtime	0.9	
modpc	1.4	no priority change
saveprcs	8.1	1 register set saved
saveprcs	12.0	all 4 register sets saved
resumprcs	21.0	
dispatch a process	22.6	
timeout and dispatch	35.7	
sendserv and dispatch	40.4	

The SEND and SIGNAL times assume that there are no processes waiting at the port or semaphore. If there is a process waiting, the time increases by approximately 5 μsec to dequeue the process and enqueue it on the dispatch port.

The operations that involve switching to a process measure the time consumed until the first instruction is executed. For instance, the operation *dispatch a process* measures the time from the point where the processor first examines the dispatch port until it executes the first instruction of the dispatched process.

MISCELLANEOUS OPERATIONS

Table 8-12 lists the times of a set of other instructions and functions, including processing of IAC messages, interrupts, and faults. All the IAC messages were sent locally to the single processor using the SYNMOVQ instruction.

The two interrupt cases measure the time from the point where the interrupt is signaled to the processor to the point where the first instruction of the interrupt handler is executed. The first case assumes that the interrupt occurs in an instruction that is suspended with resumption information being stored on the stack. The two fault cases measure the time from the execution of an ADDI instruction that overflows until the first instruction of the fault handler is executed.

REFERENCES

1.   R. P. Weicker, "Dhrystone: A Synthetic Systems Programming Bench-mark," *Comm. of the ACM*, Vol. 27, No. 10, 1984, pp. 1013-1030.

2.   C. A. Alexander *et al*, "Translation Buffer Performance in a UNIX Envi-ronment," *Computer Architecture News*, Vol. 13, No. 5, 1985, pp. 2-14.

3.   A. J. Smith, "Sequential Program Prefetching in Memory Hierarchies," *Computer*, Vol. 11, No. 12, 1978, pp. 7-21.

4.   A. J. Smith, "Cache Evaluation and the Impact of Workload Choice," *Proc. 12th Annual Symp. on Computer Architecture*, ACM, 1985, pp. 64-73.

5.   J. E. Smith and J. R. Goodman, "Instruction Cache Replacement Policies and Organizations," *IEEE Trans. on Computers*, Vol. C-34, No. 3, 1985, pp. 234-241.

Table 8-12:   Miscellaneous times (microseconds, 20 MHz)

Instruction/Operation	Time	Notes
interrupt IACM	5.9	interrupt gets posted
interrupt IACM	6.7	interrupt gets taken
test-pending-interrupt IACM	5.5	none found
preemption IACM	6.4	dispatch port is empty
flush-process IACM	31.3	
flush-TLB-physical-page IACM	26.2	nothing found
flush-TLB-physical-page IACM	30.5	something found
purge-instruction-cache IACM	5.3	
flush TLB IACM	5.4	
flush TLB PTE IACM	5.4	
freeze IACM	5.2	
synld instruction	3.6	from memory
synmov instruction	3.5	from and to memory
synmovl instruction	3.6	from and to memory
synmovq instruction	4.2	from and to memory
interrupt	3.6	with resumption information and instruction cache miss
interrupt	2.8	without resumption information
return instruction	3.5	from an interrupt handler
invoke local fault handler	5.7	
invoke supvsr fault handler	7.8	
return from fault handler	2.1	from a supervisor fault handler

APPENDIX

Instruction Summary

This appendix contains a summary of all instructions in the first-generation 80960 processors. The first column contains the instruction's assembly-language mnemonic. The second column defines the processor(s) in which the instruction appears (A for 80960KA, B for 80960KB, C for 80960MC).

The third column specifies the instruction's operands. The symbols r_a, r_b, and r_c denote operands that are registers or literals. Depending on the instruction's semantics, these can also denote floating-point registers and a set of consecutive registers. The symbol *disp* represents a displacement field value in a CTRL- or COBR-format instruction. The symbol *efa* denotes the value of the 32-bit effective address specified in a MEM-format instruction.

The fourth column summarizes the instruction's function. The function is expressed algorithmically when it can be done in only a few lines. The algorithmic description summarizes the instruction's *function*, not necessarily how the instruction is actually implemented.

ADDC	ABC	r_a,r_b,r_c	Unsigned add with carry $r_c = r_b + r_a$ + middle ccode bit ccode = 0CV where C is carryout of add and V denotes overflow if this had been a signed add
ADDI	ABC	r_a,r_b,r_c	Signed integer add $r_c = r_b + r_a$
ADDO	ABC	r_a,r_b,r_c	Unsigned add $r_c = r_b + r_a$
ADDR	BC	r_a,r_b,r_c	32/80-bit floating-point add $r_c = r_b + r_a$
ADDRL	BC	r_a,r_b,r_c	64/80-bit floating-point add $r_c = r_b + r_a$
ALTERBIT	ABC	r_a,r_b,r_c	Alter bit $r_c = r_b$ with r_ath bit assigned the value of the middle ccode bit
AND	ABC	r_a,r_b,r_c	And $r_c = r_b \ \& \ r_a$
ANDNOT	ABC	r_a,r_b,r_c	B and not A $r_c = r_b \ \& \ {\sim}r_a$
ATADD	ABC	r_a,r_b,r_c	Atomic add temp = mem(r_a & FFFFFFFC) mem(ra & FFFFFFFC) = temp + r_b r_c = temp
ATANR	BC	r_a,r_b,r_c	32/80-bit arctangent $r_c = \text{atan}(r_b/r_a)$
ATANRL	BC	r_a,r_b,r_c	64/80-bit arctangent $r_c = \text{atan}(r_b/r_a)$
ATMOD	ABC	r_a,r_b,r_c	Atomic modify temp = mem(r_a & FFFFFFFC) mem(r_a & FFFFFFFC) = ($r_c \ \& \ r_b$) | (temp & ${\sim}r_b$) r_c = temp
B	ABC	disp	Branch IP = IP + disp

BAL	ABC	disp	Branch and link $g14 = IP + 4$ $IP = IP + disp$
BALX	ABC	efa,r_c	Branch and link extended $r_c = IP + 4$ (or 8) $IP = efa$
BBC	ABC	r_a,r_b,disp	Branch if bit clear if r_ath bit in r_b clear ccode = 010 $IP = IP + disp$ else ccode = 000
BBS	ABC	r_a,r_b,disp	Branch if bit set if r_ath bit in r_b clear ccode = 010 $IP = IP + disp$ else ccode = 000
BE	ABC	disp	Branch if equal if ccode & 010 $\neq$ 0 $IP = IP + disp$
BG	ABC	disp	Branch if greater than if ccode & 001 $\neq$ 0 $IP = IP + disp$
BGE	ABC	disp	Branch if greater than or equal if ccode & 011 $\neq$ 0 $IP = IP + disp$
BL	ABC	disp	Branch if less than if ccode & 100 $\neq$ 0 $IP = IP + disp$
BLE	ABC	disp	Branch if less than or equal if ccode & 110 $\neq$ 0 $IP = IP + disp$
BNE	ABC	disp	Branch if not equal if ccode & 101 $\neq$ 0 $IP = IP + disp$
BNO	ABC	disp	Branch if not ordered if ccode = 000 $IP = IP + disp$

BO	ABC	disp	Branch if ordered if ccode & 111 $\neq$ 0 IP = IP + disp
BX	ABC	efa	Branch extended IP = efa
CALL	ABC	disp	Call procedure
CALLS	ABC	r_c	Call procedure r_c in system procedure table
CALLX	ABC	efa	Call procedure extended
CHKBIT	ABC	r_a, r_b	Store bit in condition code if r_bth bit in r_a set ccode = 010 else ccode = 000
CLASSR	BC	r_a	Determine class of 32/80-bit floating-point value
CLASSRL	BC	r_a	Determine class of 64/80-bit floating-point value
CLRBIT	ABC	r_a, r_b, r_c	Clear bit r_c = r_b with r_ath bit cleared
CMPDECI	ABC	r_a, r_b, r_c	Compare and perform signed decrement ccode = r_a ? r_b r_c = r_b - 1
CMPDECO	ABC	r_a, r_b, r_c	Compare and performed unsigned decrement ccode = r_a ? r_b r_c = r_b - 1
CMPI	ABC	r_a, r_b	Compare signed integers ccode = r_a ? r_b
CMPIBE	ABC	r_a, r_b,disp	Compare signed integers, branch if equal ccode = r_a ? r_b if ccode & 010 $\neq$ 0 IP = IP + disp
CMPIBG	ABC	r_a, r_b,disp	Compare signed integers, branch if greater than ccode = r_a ? r_b

if ccode & 001 $\neq$ 0
   IP = IP + disp

CMPIBGE	ABC	r_a,r_b,disp	Compare signed integers, branch if greater than or equal ccode = r_a ? r_b if ccode & 011 $\neq$ 0 IP = IP + disp
CMPIBL	ABC	r_a,r_b,disp	Compare signed integers, branch if less than ccode = r_a ? r_b if ccode & 100 $\neq$ 0 IP = IP + disp
CMPIBLE	ABC	r_a,r_b,disp	Compare signed integers, branch if less than or equal ccode = r_a ? r_b if ccode & 110 $\neq$ 0 IP = IP + disp
CMPIBNE	ABC	r_a,r_b,disp	Compare signed integers, branch if not equal ccode = r_a ? r_b if ccode & 101 $\neq$ 0 IP = IP + disp
CMPIBNO	ABC	r_a,r_b,disp	Compare signed integers, branch ccode = r_a ? r_b if ccode = 000 IP = IP + disp
CMPIBO	ABC	r_a,r_b,disp	Compare signed integers, do not branch ccode = r_a ? r_b if ccode & 111 $\neq$ 0 IP = IP + disp
CMPINCI	ABC	r_a,r_b,r_c	Compare and perform signed increment ccode = r_a ? r_b $r_c = r_b + 1$
CMPINCO	ABC	r_a,r_b,r_c	Compare and perform unsigned increment ccode = r_a ? r_b $r_c = r_b + 1$
CMPO	ABC	r_a,r_b	Compare unsigned integers ccode = r_a ? r_b

CMPOBE	ABC	$r_a, r_b, disp$	Compare unsigned integers, branch if equal $ccode = r_a ? r_b$ if $ccode$ & $010 \neq 0$ $IP = IP + disp$
CMPOBG	ABC	$r_a, r_b, disp$	Compare unsigned integers, branch if greater than $ccode = r_a ? r_b$ if $ccode$ & $001 \neq 0$ $IP = IP + disp$
CMPOBGE	ABC	$r_a, r_b, disp$	Compare unsigned integers, branch if greater than or equal $ccode = r_a ? r_b$ if $ccode$ & $011 \neq 0$ $IP = IP + disp$
CMPOBL	ABC	$r_a, r_b, disp$	Compare unsigned integers, branch if less than $ccode = r_a ? r_b$ if $ccode$ & $100 \neq 0$ $IP = IP + disp$
CMPOBLE	ABC	$r_a, r_b, disp$	Compare unsigned integers, branch if less than or equal $ccode = r_a ? r_b$ if $ccode$ & $110 \neq 0$ $IP = IP + disp$
CMPOBNE	ABC	$r_a, r_b, disp$	Compare unsigned integers, branch if not equal $ccode = r_a ? r_b$ if $ccode$ & $101 \neq 0$ $IP = IP + disp$

CMPOR	BC	r_a, r_b	Compare ordered 32/80-bit floating-point values
			if $(r_a = NaN) \mid (r_b = NaN)$
			    $ccode = 000$
			    raise fault if invalid_op unmasked
			else $ccode = r_a ? r_b$

CMPORL	BC	r_a, r_b	Compare ordered 64/80-bit floating-point values if $(r_a = NaN) \mid (r_b = NaN)$ $ccode = 000$

			raise fault if invalid_op unmasked else ccode = r_a ? r_b
CMPR	BC	r_a,r_b	Compare 32/80-bit floating-point values if $(r_a = NaN) \mid (r_b = NaN)$ ccode = 000 else ccode = r_a ? r_b
CMPRL	BC	r_a,r_b	Compare 64/80-bit floating-point values if $(r_a = NaN) \mid (r_b = NaN)$ ccode = 000 else ccode = r_a ? r_b
CMPSTR	C	r_a,r_b,r_c	Compare strings at addresses r_a and r_b, r_c is length

CONCMPI	BC	r_a,r_b	Conditionally compare signed integers
			if ccode ≠ 1xx
			    if $r_a \leq r_b$
			        ccode = 010
			    else ccode = 001

CONCMPO	BC	r_a,r_b	Conditionally compare unsigned integers
			if ccode ≠ 1xx
			    if $r_a \leq r_b$
			        ccode = 010
			    else ccode = 001

CONDREC	C	r_a,r_c	Conditional receive of message r_a is SS of port r_c = message SS
CONDWAIT	C	r_a	Conditional wait on semaphore r_a is SS of semaphore
COSR	BC	r_a,r_c	32/80-bit floating-point cosine r_c = cosine(r_a)
COSRL	BC	r_a,r_c	64/80-bit floating-point cosine r_c = cosine(r_a)
CPYRSRE	BC	r_a,r_b,r_c	80-bit floating-point copy reserved sign if r_b negative r_c = abs(r_a) else r_c = -abs(r_a)

CPYSRE	BC	r_a,r_b,r_c	80-bit floating-point copy sign if r_b positive $\quad r_c = abs(r_a)$ else $r_c = -abs(r_a)$
CVTILR	BC	r_a,r_c	Convert 64-bit signed integer to 32/80-bit floating-point $r_c = float(r_a)$
CVTIR	BC	r_a,r_c	Convert 32-bit signed integer to 32/80-bit floating-point $r_c = float(r_a)$
CVTRI	BC	r_a,r_c	Convert 32/80-bit floating-point to 32-bit signed integer $r_c = integer(round(r_a))$
CVTRIL	BC	r_a,r_c	Convert 32/80-bit floating-point to 64-bit signed integer $r_c = integer(round(r_a))$
CVTZRI	BC	r_a,r_c	Truncated conversion of 32/80-bit floating-point to 32-bit signed integer $r_c = integer(truncate(r_a))$
CVTZRIL	BC	r_a,r_c	Truncated conversion of 32/80-bit floating-point to 64-bit signed integer $r_c = integer(truncate(r_a))$
DADDC	BC	r_a,r_b,r_c	Decimal add with carry
DIVI	ABC	r_a,r_b,r_c	Signed integer division $r_c = r_b \div r_a$
DIVO	ABC	r_a,r_b,r_c	Unsigned integer division $r_c = r_b \div r_a$
DIVR	BC	r_a,r_b,r_c	32/80-bit floating-point division $r_c = r_b \div r_a$
DIVRL	BC	r_a,r_b,r_c	64/80-bit floating-point division $r_c = r_b \div r_a$
DMOVT	BC	r_a,r_c	Decimal move and test
DSUBC	BC	r_a,r_b,r_c	Decimal subtraction with carry

EDIV	BC	r_a,r_b,r_c	Divide 64-bit unsigned integer by 32-bit unsigned integer, producing 32-bit unsigned quotient and remainder $r_{c+1} = r_b,r_{b+1} \div r_a$ $r_c = r_b,r_{b+1} - (r_b,r_{b+1} \div r_a) \times r_a$
EMUL	BC	r_a,r_b,r_c	Unsigned integer multiply, 64-bit result $r_c,r_{c+1} = r_b \times r_a$
EXPR	BC	r_a,r_c	32/80-bit floating-point 2^x - 1 $r_c = 2^{**}r_a$ - 1
EXPRL	BC	r_a,r_c	64/80-bit floating-point 2^x - 1 $r_c = 2^{**}r_a$ - 1
EXTRACT	ABC	r_a,r_b,r_c	Extract bit field $r_c = (r_c << 32 - r_a - r_b) >> 32 - r_b$
FAULTE	ABC		Fault if equal if ccode & 010 $\neq$ 0 constraint fault
FAULTG	ABC		Fault if greater than if ccode & 001 $\neq$ 0 constraint fault
FAULTGE	ABC		Fault if greater than or equal if ccode & 011 $\neq$ 0 constraint fault
FAULTL	ABC		Fault if less than if ccode & 100 $\neq$ 0 constraint fault
FAULTLE	ABC		Fault if less than or equal if ccode & 110 $\neq$ 0 constraint fault
FAULTNE	ABC		Fault if not equal if ccode & 101 $\neq$ 0 constraint fault
FAULTNO	ABC		Fault if not ordered if ccode = 000 constraint fault

FAULTO	ABC		Fault if ordered if ccode & $111 \neq 0$ constraint fault
FILL	C	r_a,r_b,r_c	Fill string at address r_a and length r_c with value r_b
FLUSHREG	ABC		Flush all but current local register set
FMARK	ABC		Force mark (generate breakpoint trace event)
INSPACC	C	r_a,r_c	Inspect access r_c = effective rights to memory address r_a
LD	ABC	efa,r_c	Load word r_c = mem(efa)
LDA	ABC	efa,r_c	Load address r_c = efa
LDIB	ABC	efa,r_c	Load sign-extended byte r_c = mem(efa)
LDIS	ABC	efa,r_c	Load sign-extended short (halfword) r_c = mem(efa)
LDL	ABC	efa,r	Load long (double word) r_c = mem(efa) r_{c+1} = mem(efa+4)
LDPHY	C	r_a,r_c	Load physical (translated) address r_c = physical_address(r_a)
LDOB	ABC	efa,r	Load zero-extended byte r_c = mem(efa)
LDOS	ABC	efa,r	Load zero-extended short (halfword) r_c = mem(efa)
LDQ	ABC	efa,r	Load quad (four words) r_c = mem(efa) r_{c+1} = mem(efa+4) r_{c+2} = mem(efa+8) r_{c+3} = mem(efa+16)
LDT	ABC	efa,r	Load triple (three words) r_c = mem(efa)

r_{c+1} = mem(efa+4)
r_{c+2} = mem(efa+8)

LDTIME	C	r_c	Load process execution time
LOGBNR	BC	r_a,r_c	Compute 32/80-bit floating-point unbiased exponent
LOGBNRL	BC	r_a,r_c	Compute 64/80-bit floating-point unbiased exponent
LOGEPR	BC	r_a,r_b,r_c	32/80-bit floating-point logarithm $r_c = r_b \times \log_2(r_c+1)$
LOGEPRL	BC	r_a,r_b,r_c	64/80-bit floating-point logarithm $r_c = r_b \times \log_2(r_c+1)$
LOGR	BC	r_a,r_b,r_c	32/80-bit floating-point logarithm $r_c = r_b \times \log_2(r_c)$
LOGRL	BC	r_a,r_b,r_c	64/80-bit floating-point logarithm $r_c = r_b \times \log_2(r_c)$
MARK	ABC		Generate breakpoint trace event if enabled
MODAC	ABC	r_a,r_b,r_c	Read and modify (under mask) arithmetic-controls register r_c = ac ac = $(r_b$ & $r_a)$ | (ac & ~r_a)
MODPC	ABC	r_a,r_b,r_c	Read and modify (under mask) process-controls register temp = pc pc = $(r_b$ & $r_c)$ | (pc & ~r_b) r_c = temp if $r_b \neq 0$, supervisor mode is required if priority is lowered, interrupt table is checked
MODTC	ABC	r_a,r_b,r_c	Read and modify (under mask) trace-controls register r_c = tc temp = 00FF00FF & r_a tc = (temp & r_b) | (tc & ~temp)
MODI	ABC	r_a,r_b,r_c	Signed integer modulo $r_c = r_b - (r_b \div r_a) \times r_a$

$$\text{if } ((r_c \neq 0) \ \& \ (r_b \times r_a < 0))$$
$$r_c = r_c + r_a$$

MODIFY	ABC	r_a, r_b, r_c	Copy bits under mask $r_c = (r_b \ \& \ r_a) \mid (r_c \ \& \ {\sim}r_a)$
MOV	ABC	r_a, r_c	Move $r_c = r_a$
MOVL	ABC	r_a, r_c	Move long (two words) $r_c = r_a$ $r_{c+1} = r_{a+1}$
MOVQ	ABC	r_a, r_c	Move quad (four words) $r_c = r_a$ $r_{c+1} = r_{a+1}$ $r_{c+2} = r_{a+2}$ $r_{c+3} = r_{a+3}$
MOVQSTR	C	r_a, r_b, r_c	Move (copy) non-overlapping string at address r_b of length r_c to address r_a
MOVR	BC	r_a, r_c	32/80-bit floating-point move $r_c = r_a$
MOVRE	BC	r_a, r_c	80-bit floating-point move $r_c = r_a$
MOVRL	BC	r_a, r_c	64/80-bit floating-point move $r_c = r_a$
MOVSTR	C	r_a, r_b, r_c	Move (copy) possibly overlapping string at address r_b of length r_c to address r_a
MOVT	ABC	r_a, r_c	Move triple (three words) $r_c = r_a$ $r_{c+1} = r_{a+1}$ $r_{c+2} = r_{a+2}$
MULI	ABC	r_a, r_b, r_c	Signed integer multiply $rc = r_b \times ra$
MULO	ABC	r_a, r_b, r_c	Unsigned multiply $rc = r_b \times ra$
MULR	BC	r_a, r_b, r_c	32/80-bit floating-point multiply $rc = r_b \times ra$

MULRL	BC	r_a, r_b, r_c	64/80-bit floating-point multiply $rc = r_b \times ra$
NAND	ABC	r_a, r_b, r_c	Not (B and A) $r_c = \sim(r_b \ \& \ r_a)$
NOR	ABC	r_a, r_b, r_c	Not (B or A) $r_c = \sim(r_b \mid r_a)$
NOT	ABC	r_a, r_c	Not $r_c = \sim r_a$
NOTAND	ABC	r_a, r_b, r_c	Not B and A $r_c = \sim r_b \ \& \ r_a$
NOTBIT	ABC	r_a, r_b, r_c	Invert specified bit $r_c = r_b$ with r_ath bit inverted
NOTOR	ABC	r_a, r_b, r_c	Not B or A $r_c = \sim r_b \mid r_a$
OR	ABC	r_a, r_b, r_c	Or $r_c = r_b \mid r_a$
ORNOT	ABC	r_a, r_b, r_c	B or not A $r_c = r_b \mid \sim r_a$
RECEIVE	C	r_a, r_c	Receive message r_a is SS of port r_c = message SS
REMI	ABC	r_a, r_b, r_c	Signed integer remainder $r_c = r_b - (r_b \div r_a) \times r_a$
REMO	ABC	r_a, r_b, r_c	Unsigned integer remainder $r_c = r_b - (r_b \div r_a) \times r_a$
REMR	BC	r_a, r_b, r_c	32/80-bit floating-point remainder $r_c = r_b - N \times r_a$ where N is integer nearest to exact value of $r_b \div r_a$, and $abs(N) \le abs(r_b \div r_a)$
REMRL	BC	r_a, r_b, r_c	64/80-bit floating-point remainder $r_c = r_b - N \times r_a$ where N is integer nearest to exact value of

$r_b \div r_a$, and
$abs(N) \le abs(r_b \div r_a)$

RESUMPRCS	C	r_a	Resume process Discard state of current process and switch to the process whose SS is specified by r_a
RET	ABC		Return from procedure
ROTATE	ABC	r_a, r_b, r_c	Rotate $r_c = r_b$ rotated "left" by r_a positions
ROUNDR	BC	r_a, r_c	32/80-bit floating-point round $r_c = $ round_to_integral_value(r_a)
ROUNDRL	BC	r_a, r_c	64/80-bit floating-point round $r_c = $ round_to_integral_value(r_a)
SAVEPRCS	C		Save process if in process-executing state save state of current process else just flush register sets to memory
SCALER	BC	r_a, r_b, r_c	32/80-bit floating-point scale $r_c = r_b \times 2^{**}(r_a)$
SCALERL	BC	r_a, r_b, r_c	64/80-bit floating-point scale $r_c = r_b \times 2^{**}(r_a)$

SCANBIT	ABC	r_a, r_c	Scan for most-significant set bit
			if r_a contains one or more 1 bits
			    $r_c = $ position of most-significant 1 bit
			    ccode = 010
			else
			    rc = -1
			    ccode = 000

SCANBYTE	ABC	r_a, r_b	Compare bytes if any of the four pairs of corresponding bytes in r_a and r_b are equal ccode = 010 else ccode = 000
SCHEDPRCS	C	r_a	Schedule process whose SS is r_a

SEND	C	r_a, r_b, r_c	Send message r_a is SS of port r_b is message priority r_c is SS of message
SENDSERV	C	r_a	Suspend the current process and send it as a message to the port whose SS is specified by r_a
SETBIT	ABC	r_a, r_b, r_c	Set bit $r_c = r_b$ with r_ath bit set
SHLI	ABC	r_a, r_b, r_c	Signed integer left shift $r_c = r_b \ll r_a$
SHLO	ABC	r_a, r_b, r_c	Unsigned integer left shift $r_c = r_b \ll r_a$
SHRDI	ABC	r_a, r_b, r_c	Signed integer right dividing shift $r_c = r_b \div 2^{**}(r_a)$
SHRI	ABC	r_a, r_b, r_c	Signed integer right shift $r_c = r_b \gg r_a$
SHRO	ABC	r_a, r_b, r_c	Unsigned integer right shift $r_c = r_b \gg r_a$
SIGNAL	C	r_a	Signal semaphore r_a is SS of semaphore
SINR	BC	r_a, r_c	32/80-bit floating-point sine $r_c = \text{sine}(r_a)$
SINRL	BC	r_a, r_c	64/80-bit floating-point sine $r_c = \text{sine}(r_a)$
SPANBIT	ABC	r_a, r_c	Span most-significant set bits (scan for most-significant clear bit if r_a contains one or more 0 bits $\quad r_c$ = position of most-significant 0 bit $\quad$ ccode = 010 else $\quad$ rc = -1 $\quad$ ccode = 000
SQRTR	BC	r_a, r_c	32/80-bit floating-point square root $r_c = \text{sqrt}(r_a)$

SQRTRL		BC	r_a,r_c	32/80-bit floating-point square root

$r_c = \text{sqrt}(r_a)$

ST	ABC		r_c,efa	Store word

$\text{mem}(efa) = r_c$

STIB	ABC		r_c,efa	Store integer byte

$\text{mem}(efa)$ = low-order byte of r_c
can result in overflow

STIS	ABC		r_c,efa	Store integer short (halfword)

$\text{mem}(efa)$ = low-order halfword of r_c
can result in overflow

STL	ABC		r_c,efa	Store long (double word)

$\text{mem}(efa) = r_c$
$\text{mem}(efa+4) = r_{c+1}$

STOB	ABC		r_c,efa	Store byte

$\text{mem}(efa)$ = low-order byte of r_c

STOS	ABC		r_c,efa	Store short (halfword)

$\text{mem}(efa)$ = low-order halfword of r_c

STQ	ABC		r_c,efa	Store quad (four words)

$\text{mem}(efa) = r_c$
$\text{mem}(efa+4) = r_{c+1}$
$\text{mem}(efa+8) = r_{c+2}$
$\text{mem}(efa+12) = r_{c+3}$

STT	ABC		r_c,efa	Store triple (three words)

$\text{mem}(efa) = r_c$
$\text{mem}(efa+4) = r_{c+1}$
$\text{mem}(efa+8) = r_{c+2}$

SUBC	ABC		r_a,r_b,r_c	Unsigned subtract with carry

$r_c = r_b - r_a - 1 + \text{middle ccode bit}$
ccode = 0CV
where,
C is carryout of $r_b + \sim r_a + \text{middle ccode bit}$
V denotes overflow from this addition

SUBI	ABC		r_a,r_b,r_c	Signed integer subtract

$r_c = r_b - r_a$

SUBO	ABC	r_a, r_b, r_c	Unsigned integer subtract $r_c = r_b - r_a$
SUBR	BC	r_a, r_b, r_c	32/80-bit floating-point subtract $r_c = r_b - r_a$
SUBRL	BC	r_a, r_b, r_c	64/80-bit floating-point subtract $r_c = r_b - r_a$
SYNCF	ABC		Synchronize faults if ~nif wait until no imprecise faults can be generated by any uncompleted instructions

SYNLD	ABC	r_a, r_c	Synchronous load
			if address translation enabled
			    temp = physical_address(r_a)
			else temp = r_a
			temp = temp & FFFFFFFC
			if temp = FF000004
			    r_c = interrupt_control_register
			    ccode = 010
			else
			    r_c = mem(temp)
			    if bad access
			        ccode = 000
			    else ccode = 010

SYNMOV	ABC	r_a, r_b	Synchronous move
			if address translation enabled
			    temp = physical_address(r_a)
			else temp = r_a
			temp = temp & FFFFFFFC
			if temp = FF000004
			    interrupt_control_register = mem(r_b)
			    ccode = 010
			else
			    temp1 = mem(r_b)
			    mem(temp) = temp1
			    if bad access
			        ccode = 000
			    else ccode = 010

SYNMOVL	ABC	r_a, r_b	Synchronous move long if address translation enabled temp = physical_address(r_a) else temp = r_a temp = temp & FFFFFFF8

temp1 = mem(r_b)
temp2 = mem(r_b + 4)
mem(temp) = temp1
mem(temp + 4) = temp2
if bad access
   ccode = 000
else ccode = 010

SYNMOVQ	ABC	r_a,r_b	Synchronous move quad

if address translation enabled
   temp = physical_address(r_a)
else temp = r_a
temp = temp & FFFFFFF0
temp1 = mem(r_b)
temp2 = mem(r_b + 4)
temp3 = mem(r_b + 8)
temp4 = mem(r_b + 12)
if temp = FF000010
   ccode = 010
   use temp1..temp4 as received IAC msg
else
   mem(temp) = temp1
   mem(temp + 4) = temp2
   mem(temp + 8) = temp3
   mem(temp + 12) = temp4
   if bad access
      ccode = 000
   else ccode = 010

TANR	BC	r_a,r_c	32/80-bit floating-point tangent

r_c = tangent(r_a)

TANRL	BC	r_a,r_c	64/80-bit floating-point tangent

r_c = tangent(r_a)

TESTE	ABC	r_a	Test for equal

if ccode & 010 $\neq$ 0
   r_a = 1
else r_a = 0

TESTG	ABC	r_a	Test for greater than

if ccode & 001 $\neq$ 0
   r_a = 1
else r_a = 0

TESTGE	ABC	r_a	Test for greater than or equal

if ccode & 011 $\neq$ 0

$r_a = 1$
else $r_a = 0$

TESTL	ABC	r_a	Test for less than if ccode & 100 $\neq$ 0 $r_a = 1$ else $r_a = 0$
TESTLE	ABC	r_a	Test for less than or equal if ccode & 110 $\neq$ 0 $r_a = 1$ else $r_a = 0$
TESTNE	ABC	r_a	Test for not equal if ccode & 101 $\neq$ 0 $r_a = 1$ else $r_a = 0$
TESTNO	ABC	r_a	Test for not ordered if ccode = 000 $r_a = 1$ else $r_a = 0$
TESTO	ABC	r_a	Test for ordered if ccode & 111 $\neq$ 0 $r_a = 1$ else $r_a = 0$
WAIT	C	r_a	Wait on semaphore r_a is SS of semaphore
XNOR	ABC	r_a, r_b, r_c	Exclusive nor $r_c = \sim(r_b \oplus r_a)$
XOR	ABC	r_a, r_b, r_c	Exclusive or $r_c = r_b \oplus r_a$

Index